HOW THE
GREAT
PYRAMID
WAS BUILT

HOW THE GREAT PYRAMID WAS BUILT

Craig B. Smith

Foreword by Zahi Hawass
Photography by Andy Ryan

SMITHSONIAN BOOKS
WASHINGTON

Citing the company names DMJM, The Ralph M. Parsons Company, and Bechtel Corporation:
Permission granted by each company.

Excerpt from Carl Sandburg, "Cool Tombs," in *Cornhuskers*, by Carl Sandburg, © 1918 by Holt, Rinehart and
Winston and renewed 1946 by Carl Sandburg, reprinted by permission of Harcourt, Inc.

Excerpt from Andrew Marvell, "To His Coy Mistress," in *Six Centuries of Great Poetry*, by Robert Penn Warren and
Albert Erskine, © 1955, reprinted by permission of Random House.

Executive editor: Scott Mahler

Copy editor: Tom Ireland

Production editor: Joanne Reams

Designer: Michael Hentges

Library of Congress Cataloging-in-Publication Data
Smith, Craig B.
   How the Great Pyramid was built / Craig B. Smith ; foreword by Zahi Hawass.
     p. cm.
   Includes bibliographical references and index.
   ISBN 1-58834-200-X (cloth: alk. paper)
   1. Great Pyramid (Egypt) 2. Pyramids—Egypt—Design and construction. I. Title.

   DT63.S6 2004
   932—dc22   2004054651

British Library Cataloging-in-Publication Data are available

Manufactured in the United States of America

09 08 07 06 05 04   1 2 3 4 5

♾ The paper used in this publication meets the minimum requirements of the American Standard for
Information Sciences—Permanence of Paper for Printed Library materials ANSI Z9.48-1992.

For permission to reproduce illustrations appearing in this book, please correspond directly with the owner
of the works as listed in the illustration credits on page 285. Smithsonian Books does not retain reproduc-
tion rights for these illustrations individually or maintain a file of addresses for photo sources.

CONTENTS

Foreword by Zahi Hawass 7

Author's Note 10

Introduction: Quest for Answers 16

1. Egypt during the Old Kingdom 28

2. Pyramid Evolution 58

3. A Tomb for a King 86

4. We Consecrate This Holy Ground 108

5. The Desert Comes Alive 126

6. Soaring toward the Heavens 150

7. Stair Steps to the Gods 178

8. The Workforce 202

9. The Construction Schedules 222

10. Life Everlasting 234

Appendixes

1. Egyptian Gods 245

2. Units of Measurement and Other Technical Data 248

3. Calculating the Number of Blocks in the Pyramid 250

4. A Primer on Program Management 252

Notes 262

Glossary 272

Annotated Bibliography 277

Illustration Credits 285

Index 286

FOREWORD

The pyramids have always captured our hearts. Magic and mystery surround them, filling scholars and the public with wonder that evokes the following question: How did our ancestors, 4,500 years ago, build these magnificent structures, and how did they organize the workforce that was necessary for their construction?

Our excavations in Giza have helped to shed some light on the mystery of the pyramids. We discovered the location of the quarry for the Great Pyramid of Khufu on the south side as well as the area of the ramps they used. We also learned how they constructed the base. The discovery of the tombs of the pyramid builders was a significant find that has provided us with important information about the workmen who dragged the stones, the artisans who sculpted statues and supervised construction of the pyramid base, the people who worked at the harbor, the overseer of the craftsmen, and the various people ("overseer of the side of the pyramid," "director of the draftsmen," "overseer of masonry," "director of workers," etc.) who were involved in the construction of the pyramids.

The sand and stones of Egypt hold the mystery of its past, and through our work it has been amazing and extremely rewarding to see some of these secrets revealed. Scholars always believed that the pyramid builders ate bread, garlic, and onion, but recent discoveries show that they slaughtered eleven cattle and thirty-three sheep and goats every day. They had enough meat to feed 10,000 workers a day. It is important to know that these young people, who worked from sunrise to sunset, were well fed.

The workers were organized into crews and then further divided into gangs. Each gang had an overseer. Every gang had a name such as Friends of Khufu. The gangs were divided into phyles (a Greek word meaning "tribes"). Each phyle consisted of 200 workmen and also had overseers. The phyles had such names as the Great Ones and the Green Ones. Each phyle was divided into units that consisted of about ten to twelve workers, and these units had such names as Perfection. I believe that the extensive organization involved in building the Great Pyramid is more important than the pyramid itself. We engineers, architects, and scholars could not succeed in building the Great Pyramid of Khufu today because we do not have the same dedication as the ancient Egyptians. The building of the pyramids was the national project of Egypt. Every household in Upper and Lower Egypt participated in the building of the king's pyramid by sending workforces, food, and supplies. They did this to help the king be a god; the pyramid was the ladder that the king would climb to the afterlife on his quest for immortality.

I used to teach a summer course at UCLA, and during that time my friend Elizabeth Brooks introduced me to Craig Smith. He was interested in the construction and management of building the pyramids. His background was in architecture and engineering, but he was as interested in the people who built the pyramids as he was in the engineering and construction. Elizabeth Brooks organized a conference on the pyramids at UCLA, and Craig Smith's lecture impressed me. He subsequently came to visit me in Egypt and conducted his own research, which entailed visiting many important sites, among them the tombs of the pyramid builders and our other archaeological sites at Giza.

How the Great Pyramid Was Built is a significant book for students, scholars, and the general public. The author explains the history and the development of the pyramids from the mastaba to the Step Pyramid to a true pyramid. Craig Smith was motivated to write this book because of his engineering background, and he has accomplished something quite remarkable. By looking at the Great Pyramid as a straightforward construction project—something never done before to this level of detail— he has independently confirmed what archaeologists have recently come to believe about the size of the workforce. The difference is, we couldn't prove it.

Through the years, many books have been written about pyramids. Scholars continue to excavate around the pyramids and publish their discoveries. Every discovery reveals more information about the pyramids and the incredible people who constructed them.

The mystery of the Great Pyramid of Khufu remains; however, we feel that we are making progress. In 1993 the so-called door (20 centimeters square), inside the south shaft of the so-called Queen's Chamber, was found with two copper handles. Our exploration in September 2002 revealed another door behind the first and another in the northern shaft. We ask ourselves, "Are these doors for symbolic use, for the king to reach the other world safely as a part of his quest for immortality, or are they hiding a great discovery?" We hope to have our answers soon. We will continue to explore, excavate, and research until we reveal all the mysteries of the Great Pyramid of Khufu.

—Zahi Hawass

Dr. Hawass is the secretary general of the Supreme Council of Antiquities and the director of excavations at Giza, Saqqara, and Bahariya Oasis.

AUTHOR'S NOTE

A few words about the preparation of this book are in order. First, I am not an archaeologist. I've drawn heavily on the knowledge and expertise of two leading authorities—Zahi Hawass and Mark Lehner—for anything related to archaeology. They are considered among the best in their field. My interest lies in the engineering and construction of the Great Pyramid, and that coincides with my own experience and expertise. Since a substantial part of the pyramid is still standing—albeit in a slightly damaged condition—it is possible to examine the construction and determine how some tasks were undertaken and executed. The same is true of the architecture and engineering design.

This type of "forensic analysis" can only be carried so far, however; then one encounters a gap in solid information and findings. I decided to investigate this gap with the modern tools available to someone who plans and manages large public works projects. By means of this approach, bits and pieces of factual data can be assembled, and models can be developed that lead to a reasonably good understanding of how the project was implemented. Certain assumptions must also be made— for example, how many loads of soil or sand could an ancient Egyptian laborer haul per day? Fortunately, these things can be estimated with a good degree of accuracy. When we are less certain, a technique called "sensitivity analysis" can be used to check how sensitive the final result is to the analysis. If large changes in the estimate do not change the final outcome, then the model is said to be insensitive to that particular parameter. For example, if the labor in hauling soil and sand was only 2 percent of the labor involved in cutting and finishing blocks of limestone, then it wouldn't matter very much if we were off by 10 percent or even 50 percent in our estimate of sand-hauling labor. I endeavored to make these analyses as I checked the data and the assumptions I used.

Finally, as skilled as the ancient Egyptians were in constructing magnificent monuments that survived forty-five centuries of human abuse and the wear and tear attributed to the elements, they were apparently far less concerned with leaving a paper trail of documents describing their plans and decisions. Most of the written records that have survived were retrieved from graves and tombs. From these we have been able to learn about the day-to-day lives of the workers and high officials and study the tools and methods they used, and from this, infer other things.

Very little is directly known about the key players in the Great Pyramid drama—about their families, jobs, daily routines, frustrations, challenges. But based on my personal experience with large public works projects, I know that every day brings a little drama. Things don't go well, designs have errors, materials do not meet specifications or are not delivered on time, schedules slip, accidents happen, budget problems arise. Communication and coordination are frequently cited as the major causes of problems. The need to resolve these issues is integral to every complex project. These same challenges undoubtedly faced the builders of the Great Pyramid. The details make the story more interesting, lend a degree of realism, and make the ancient Egyptians' accomplishments all the more impressive.

Consequently you will find some speculation in this book concerning problems that may have arisen and alternative methods the builders may have employed to solve these problems. I have also tried to resurrect the personalities of the principal players behind the Great Pyramid because such a tremendous accomplishment could never have happened without strong-willed, talented, determined people driving the project. These individuals bring the story to life for the reader who does not typically follow heavy construction, and their stories clarify some of the more abstract technical issues that might otherwise be boring. My research reveals events that must have occurred. We know, for example, that the pharaoh "stretched the cord" on major construction projects, that the vizier had oversight of major projects, that last-minute changes to the design of Khufu's pyramid were necessary, and that there were quality-control problems.

I've drawn on my experience with hundreds of projects in which I participated as an engineer, builder, or construction executive, noting how similar but lesser projects were managed and constructed despite the inevitable obstacles that arose. The ancient dates in the text are derived from my schedule analysis and are predicated on the assumption that after his first year as pharaoh, Khufu ordered the work on his pyramid to begin.

The Third and Fourth Dynasties are those reported in Lehner's *The Complete Pyramids*. None of these dates are known with absolute certainty because there is still some debate about when Khufu's reign began and how long it lasted. As for the general time frame, careful carbon dating of materials and artifacts extracted from the Great Pyramid places its origin at around 2694 BC—a hundred-plus years older than the dates I have used. Since the uncertainty in such measurements could easily be plus or minus 100 years (if the wood that was the carbon source came from 100-year-old trees, for example), a mere century one way or the other seems inconsequential.

The technical facts are drawn as much as possible from my direct observations and measurements made during my several visits to Giza, Saqqara, and Dahshur. When this was not possible, I consulted other experts or specialists and reviewed the extensive technical literature. The bibliography lists these sources. I built models to test the ramps and other concepts described in this book, made copper tools to see how well they worked, and constructed replicas of Egyptian measuring instruments to see how accurate they could be.

To give a brief synopsis of my qualifications to undertake this work, early in my career as an engineer I specialized in seismic and vibration problems of large structures and became familiar with high-rise buildings, dams, commercial nuclear power plants, and other forms of heavy construction. As a design-builder, I oversaw the design and construction of buildings, industrial facilities, test laboratories, and a waste-to-energy power plant. As a construction executive, I participated in program and construction management projects involving the construction of large rapid transit systems, the rebuilding of university buildings damaged by a major earthquake, the expansion of various airports, and the renovation of the Pentagon, which is still under way. Some of these projects were in distant lands. I helped build small factories in the hinterlands of Brazil, worked on subways in Korea, and was involved in projects in Asia, Europe, and the Middle East.

I am indebted to a number of people who helped make this book a reality. First, without the encouragement of Zahi Hawass and Mark Lehner, I would not have undertaken this project. They shared ideas, invited me to their research sites, and patiently answered questions. I am especially indebted to Dr. Hawass, who provided access to any pyramid site I chose to visit. He is a wonderful ambassador for Egypt and its spectacular accomplishments, both ancient and modern. I am grateful to the talented staff at the Cairo Marble and Granite Company, who showed me current techniques for cutting limestone and granite and who facilitated my visit to the modern quarries.

The support of my colleagues at Daniel, Mann, Johnson & Mendenhall, and Holmes and Narver, Inc., is gratefully acknowledged. In particular I thank Linda Barhouse, Ernie Bustamante, Donna Cant, Patricia Ferguson, Clyde Garrison, Johnson Nee, Allyn Simmons, and Alexandra Spencer for their assistance. Ray Holdsworth, president, AECOM Technology Corporation, provided support and encouragement. Stuart Ockman, of Ockman & Borden Associates, one of the world's foremost authorities on critical path scheduling, kindly reviewed the schedule and rendered valuable assistance.

My agent, Ron Goldfarb, believed in the book and helped it become a reality. I am deeply indebted to Scott Mahler and the excellent staff of Smithsonian Books who produced the book.

Finally, words alone do not do justice to the dedication, guidance, and insights provided by Andy Ryan (photography), Kurt Mueller (illustrations), Anne Elizabeth Powell (manuscript review), and Nancy J. Smith (research assistance and critiques).

HOW THE GREAT PYRAMID WAS BUILT

INTRODUCTION
Quest for Answers

GIZA, EGYPT (NOVEMBER 1981)

We step from the bus into the dusty parking lot. Dozens of Arab vendors are milling around, and as we trickle into their midst they begin pressing close to us, trying to sell us cheap beaded necklaces and *al-ghuṭra* and *al-ʿiqāl*—the traditional Arab headdress. Then the camel drivers spot us and descend in droves, hounding us to ride their camels for a price, their badgering relentless. People on horseback wind their way through the confusion—some are vendors, some uniformed police—and it's hard to imagine how we will ever make our way through this rabble.

Our cruise ship is docked in Port Said, where we boarded buses that brought us first to Cairo and now to Giza. When we left the port it was cool, but now we feel the full impact of the desert sun, and the heat thickens the smell engulfing us—a potent blend of crowded humanity and animal flesh laced with traces of camel and horse dung. Above all, the din assaults us, an incessant, incongruous cacophony so at odds with the intrinsic sanctity of the place.

I pull away from the jostling crowd to strike off on my own. With only three hours before we must reboard the bus for the long trip back to Port Said, I want as much time to myself up there on the hill as I can squeeze from our stop. The Great Pyramid at Giza, the tomb of the Egyptian pharaoh Khufu and the sole survivor of the Seven Wonders of the Ancient World, commands the plateau, looming above and before me, but I'm intent upon seeing it up close, touching it, studying it, digesting it.

As I make my way up the hill, the sun is now directly overhead, and in the brilliance of this light the pyramid seizes my imagination, a structural colossus vastly more extraordinary than I had imagined—*staggering* is the word for it. As an engineer and constructor, I have long been fascinated with how ancient civilizations constructed buildings, towns, and roads with limited tools and technology but incredible skill—how they moved massive blocks of stone, erected immense structures without mortar, finished joints so fine that their crevices would not allow the intrusion of even a razor blade. The roads and aqueducts of the Romans, the temples of the Greeks, the ruins of the Mayans and Peruvians—all places I have visited in the course of my work on one project or another—intrigue me, as they have intrigued humankind for millennia.

But no monument has captivated the world more than the Great Pyramid. The most celebrated and enduring vestige of the remarkable civilization that emerged during the golden age of the pharaohs, it is a structure without equal, a singular landmark within the human experience. Roughly two-thirds the size of Hoover Dam, it rises 146.6 meters on a base covering 5.3 hectares, incorporates 2.6 million cubic meters of material (an estimated 2.3 million blocks of limestone and granite), and for more than 4,300 years stood as the tallest structure on earth—until the Washington Monument eclipsed it at 169 meters in 1885.

Yet it is astonishing not just for its size. Architecturally it is stunning—a testament to the ancient Egyptians' prodigious grasp of mathematics and their mastery of spatial concepts. It is oriented almost precisely true north and south; the angle of inclination of the triangular faces is accurately maintained at $51.9°$. The base is level to less than 2 centimeters, the average deviation of the faces from the cardinal directions is $3'6''$ of arc, and the greatest difference in the length of the faces from the average length of 23,040 centimeters ranges from 1 to 11 centimeters (equivalent to a variation of 0.4 to 4.3 inches in 756 feet). Also, whether by coincidence or intent, the area of a triangular face is equal to the square of the height. In short, the precision of its configuration is phenomenal. In the shadow of this magnificent structure, I am humbled by the brilliance of its builders.

It is a structure that surpasses belief, and no one who experiences it firsthand can fail to be awed by it. In my mind's eye, I see it as it stood when completed in 2540 BC—faced with white limestone, dazzling in the sunlight. Then I try to imagine it by moonlight, and this reverie fills me with a longing to forget about the waiting bus, sneak around to the cemetery behind the pyramid, find a vacant tomb, and wait there for the tourists to disappear, the sun to go down, and quiet to descend upon the Giza Plateau. Alone in the stillness of dusk, perhaps I could sense the presence of the ancients who created this place, and perhaps then I could fathom why they selected this site and what they sought to accomplish by erecting on it this tomb for their pharaoh—this enormous, perfect pyramid.

But I don't sneak away. There are two other noteworthy pyramids to be explored here, and I hurry across the plateau to examine them as well. Individually and collectively, the Giza pyramids are remarkable. Not only are they the largest and best preserved of the nearly 100 pyramids still standing on the west bank of the Nile, but they also are aligned with exceptional precision. The southeast corners of each pyramid are positioned almost precisely on a diagonal that extends 43° true north, creating a spectacular conjunction of shadows—this despite the fact that the compass had not yet been invented.

The pyramids at Giza represent three generations of pharaonic rule: Khufu; his second-reigning son, Khafre; and his grandson, Menkaure. Khufu's pyramid is the largest of the three principal pyramids at Giza. Next to it, and only slightly smaller, is the pyramid constructed by Khafre. The third and smallest of the three large pyramids was built by Menkaure. The pyramids of Khafre and Menkaure retain some of the casing stones that once gave them their perfect pyramidal shape—the type of stones stolen from Khufu's pyramid through the ages. On this particular day, only Menkaure's pyramid is open to tourists; though disappointed that I won't see the interior of Khufu's great tomb, I am grateful for the opportunity to investigate any of the three.

And so once more I am part of a crowd, hunched over, crab-walking down the Descending Corridor into the recesses of Menkaure's pyramid, listening to our Arab guide explain the ancient Egyptian burial process. A woman behind me giggles nervously—feeling a bit claustrophobic, perhaps—as we move deeper into the bedrock, millions of pounds of stone stacked overhead. "Do they have earthquakes here?" she asks of no one in particular.

Below ground, the air—fresh and cool at first—becomes heavy, warm, slightly foul, and musty, as the air of an ancient tomb might feel and smell, and it makes the transcendence of more than four millennia tangible, real. What an extraordinary privilege we are being afforded by our descent into this tomb—the pyramids were never meant to be explored. They were the burial places of divine beings, as the ancient Egyptians deemed their pharaohs to be, and as such were sanctified and inviolable—far removed from the domain of mortals.

I am struck by the fervid dedication that would have enabled a people to perform this backbreaking, mind-numbing labor—that would have prompted them to dig deep into the bedrock of this plateau with only crude tools at their disposal. These were people who had no pulleys, no wheels, no iron tools, no compasses—just chisels, saws, hammers, and drills made of copper, wood, and stone—and yet they flawlessly designed, sited, and erected structures of precise geometric configuration and complex construction on an astonishing scale.

The Descending Corridor is dimly lit, augmenting the eerie feeling that arises from entering this ancient tomb. My hand glides along the corridor wall, feeling tool marks 4,000 years old, the only surviving relics of anonymous stonemasons, who never could have imagined that elements of their work would be remarked upon by future generations through the ages. I examine the chisel marks, trying to calculate the number of strokes they represent, and imagine the sheer exhaustion of those who wielded those chisels as they labored hour upon hour, breathing fetid air, covered in dust, their work illuminated by the dim light of a candle or oil-wick lamp. Equally weary would be those who hauled tons of stone by the basketful from the tunnel, much of it reduced to chips.

After some fifty steps, we enter a chamber embellished with a series of carved doors. Beyond it is a portcullis, where at one time three large stone doors blocked the entrance to the tomb. A horizontal passage leads into a second chamber—larger than the first—that has a narrow tunnel leading down into the actual burial chamber. The burial chamber is lined with granite and has a vaulted roof. The stonework is exquisite, given the hardness of granite and the tools available to the ancient Egyptian workers. Even though it has been defaced by graffiti through the years, it is still remarkable. I cannot imagine how the builders maneuvered the heavy granite beams down the steep, narrow tunnels to complete this room.

There is no sarcophagus. The guide explains, "It was very beautiful, but it was stolen by the British. They put it on a ship bound for England. Unfortunately, the ship sank, and the sarcophagus was lost forever. Some people think that this proves the power of the ancient gods. Remember this if you are tempted to remove any artifacts from Giza!"

When we climb back up to the daylight, the afternoon is slipping away. The shadow of Khufu's pyramid is beginning to creep across the plateau with the movement of the sun, telling us time's up, we must go. But I am held utterly spellbound by this place, conquered by the pyramid's power. What must it have been like to know that you could create such magnificence by human labor alone?

I long to be able to climb Khufu's pyramid and examine it closely. Study the size and placement of the stones, the slope of the structure. See precisely how it rises—ton upon ton for hundreds of meters. Analyze its construction up close. Apply more than twenty years of engineering and construction experience to it and see what I find. I want to know everything there is to know about this monument—and about the people who built it. The majesty of the ancient Egyptians' thinking and the magnificence of their achievements are quite simply breathtaking. Could we have accomplished what they did without modern technology and equipment? Could we have done it without the pulley, the compass, or the wheel?

I am among the last to board the bus, and as it pulls away, I know that I will return. I simply must find out how these structures were built. During the bus ride, my mind keeps asking: how? How?

At the time, I never imagined that the answer was not already known or that I would launch my own exhaustive investigation into the construction of this, the largest pyramid. I just knew that somehow, some way, I had to discover how and why the ancient Egyptians built these pyramids.

LOS ANGELES, CALIFORNIA (1982–1992)

During the decade that passes, my research file on the Giza pyramids grows—as do my job responsibilities. In 1988 I leave the engineering consulting firm I founded in 1972 to become the president of a local architecture, engineering, and construction firm. In 1992 I accept the position of vice president, construction and facilities management, for DMJM, a large national architecture, engineering, and construction firm with global operations. My children graduate from college, find jobs, launch independent lives. In other words, life progresses, and career demands leave me precious little time for any serious investigative work on the building of the pyramids.

Shortly after returning home to California from Egypt in 1981, I discover that little is known about how the Giza pyramids were built. There are theories, of course, but no detailed investigation from a construction perspective. My thoughts and interest focus on Khufu's pyramid—the largest, most complex, and by far the most spectacular. If I can unravel the secrets of Khufu's pyramid, there will be less mystery regarding the others. In what little spare time I have, I collect all the information I can on the Great Pyramid in the hope of analyzing how it was constructed, but in fact I accomplish very little because I have yet to hit upon a suitable approach to such an analysis.

OAKLAND, CALIFORNIA (NOVEMBER 1993)

It is raining, about 10:30 p.m., the end of the day in a seemingly endless string of days. I am in the Hilton Hotel at the Oakland airport. I have been here for nearly two weeks, working twelve-hour days, and I'm getting tired. I've run out of clean clothes and am running out of money. I need a break to find a laundry and a bank. I wonder if I can get home to my family for Thanksgiving. My job here is to lead the mobilization effort for a new program management job for a Northern California rapid transit agency. The project involves the extension of a transit system, including the construction of two major stations, parking facilities for 3,000 cars, and the construction of 14 miles of trackway and several bridges. A good portion of the electrified rail system is being constructed down the median of a major freeway, a heavily traveled commuter route that takes workers into Northern California's Silicon Valley. This job had been in progress for about a year under a previous program manager and is in trouble. It's behind schedule and over budget. There have been several accidents—one that included the toppling of a large crane. A woman was recently killed in a collision on the freeway involving one of the construction vehicles. After this occurred, DMJM was brought in to solve the problems and get the project back on track, and to do so, I am putting in place an entirely new team of construction managers, estimators, schedulers, and inspectors—setting up a completely new organization while ensuring that the work continues without a hitch.

The project involves twelve major construction contracts and hundreds of subcontractors. The overall cost will exceed half a billion dollars and involve the excavation of nearly 138,000 cubic meters of soil and rock and the placement of more than 46,000 cubic meters of concrete. It is a four-year project expected to take nearly 1,000 person-years of labor to complete. As the principal in charge, it is my responsibility to make sure that we provide the project team with all of the resources and tools they need to complete the job satisfactorily.

A friend has lent me I. E. S. Edwards's book, *The Pyramids of Egypt*.[1] As I sit pondering the transit project, I spot the book on the bedside table. My mind shifts gears. How, I wonder, did the ancient Egyptians deal with these same problems? The challenge of assembling a skilled workforce, preparing the site, bringing the materials to the site in a timely manner, providing for all the logistical needs—roads, a harbor, food, housing? They had no computers or any of the other tools we as modern program managers use routinely on massive public works projects.

Suddenly, it hits me: I know how I am going to analyze the construction of the Great Pyramid. My research thus far has given me a good grasp of the technology that was available to the ancient Egyptians; what has been missing is an analytical framework on which to superimpose what I know about the pyramid's construction, what is not known, and what I now believe is possible in terms of methods and techniques. In this moment of insight (brought about by exhaustion!), it occurs to me that the framework I need can be constructed with the same tools we use today in managing any large public works project—the tools I am using at this very moment. These tools take complex construction projects and break them down into minute individual operations or steps using a process called a "work breakdown structure." Once this has been done, the resources needed to perform the project—materials and labor—can be determined accurately, and the schedule to complete the work can be established. I can construct such a framework for the construction of the Great Pyramid by identifying and analyzing each of the steps it took to build the pyramid—from selecting and preparing the site to placing the last of the white casing stones on the pyramid face.

The more I think about this approach, the more convinced I become that it would provide new insights into how the Great Pyramid was constructed. My experience tells me that it could not have been built without some form of organized program management. The logistical challenges of executing this enormous undertaking in an inhospitable desert were simply too great for the project to have been undertaken spontaneously, or by volunteers, or even by slaves. No, there had to have been a plan—a complex, well-thought-out plan—and there had to have been someone behind the plan, a master builder in charge of the work. My research suggests this person must have been Hemiunu, Pharaoh Khufu's vizier, a man who also held the title of overseer of all the king's works.

Hemiunu was Khufu's cousin and clearly occupied a position of great importance during Khufu's reign. His tomb was discovered in 1912, in the western cemetery behind the pyramid. Little is known about him, but what is known reflects his importance. His tomb is one of the largest in the cemetery and, when opened, was found to contain a life-size statue of him. The head was broken, damaged by grave robbers who had gouged out the precious stones or metal used to form the statue's eyes. Today, the statue resides in the Roemer and Pelizaeus Museum, in Hildesheim, Germany. It is the statue of a middle-aged man, a little on the stout side, with the confident face of one accustomed to wielding authority, a man who could conceive an enormous work and see it through to completion.

I continue to think about how the Great Pyramid was constructed and conduct research in my spare time. To motivate myself, I agree to present a paper on the topic at a national technical conference. I try to imagine how Hemiunu would have approached certain issues and solved logistical problems, given the resources that he did or did not have. I imagine conversations; I hear him giving orders. I try to place myself on the site—try to feel the heat; to envision the dust, the thousands of workers going about their daily tasks, the smoke from forges and cooking fires; to hear the din of masons pounding on limestone.

Facing the impending deadline for the conference, I assemble a group of DMJM's construction experts in a brainstorming session to examine various aspects of the pyramid's construction and seek their opinions on what tasks were necessary and how the ancient Egyptians might have performed them. I build computer models to help me study the construction process, incorporating all of the data gathered up to this moment.

Several weeks later I have the first solid results—and am surprised that my findings seem to fly in the face of conventional wisdom. The results indicate that it took less time and far fewer workers than previously thought to complete the construction. Rather than run the risk of presenting my findings and embarrassing myself publicly, I seek out the two most preeminent Egyptologists working at the Giza site and ask them to review my work. They are Zahi Hawass, an Egyptian, who is director of excavations at Giza, Saqqara, and Bahariya Oasis; and Mark Lehner, an American who is the visiting assistant professor of Egyptian archaeology at the Oriental Institute of the University of Chicago and a research associate at the Harvard Semitic Museum. They tell me that my findings corroborate what they are learning from field archaeology, and they are intrigued that I have reached my conclusions from a program/construction management approach. They invite me to visit them in Egypt.

Relieved and heartened by their encouragement, I present my preliminary findings on the construction of Khufu's pyramid at the annual meeting of the Construction Management Association of America in October 1996 in Baltimore. The interest it generates is astonishing. Letters and e-mails arrive from around the world, and I am asked to take part in an Arts and Entertainment Channel (A&E) television documentary titled *The Great Builders of Egypt*.[2] In February 1997 I return to Egypt to meet with Hawass and Lehner and learn about their work in detail, conduct fieldwork, gather data, and study not only the pyramids at Giza but also their predecessors at Saqqara and Dahshur. Photographer Andy Ryan joins me there and takes an initial set of superb photographs.

In February 1999 *Great Builders of Egypt* airs on A&E, and as a result *Civil Engineering* magazine expresses interest in publishing the paper I presented in Baltimore as a cover story.[3] Would I be willing to work with them on this? I would indeed. The article, published in the June 1999 issue, generates even more letters and e-mails from around the world than did the original paper. As a consequence I am included in a Public Broadcasting System documentary, *Lost Cities of the Pyramids*, which also features Hawass and Lehner[4]

In April 2000 Ryan and I return to Giza. I have decided to write a book on my findings, and with permission granted by Hawass, we are given broad access to the monuments, enabling me to conduct a more detailed investigation and Ryan to translate my work onto film. We explore not only the pyramids at Giza but those at Saqqara, Dahshur, and Meidum as well—more than a dozen in all.

This trip is particularly significant because now I begin to grasp the evolution of the building technology that made the construction of the Great Pyramid possible. The earlier constructions of the Step Pyramid, then the Bent Pyramid, and finally the Red (North) Pyramid were precursors to the construction of Khufu's pyramid. New ideas were tested in full-scale prototypes. Some ideas worked, some failed. The best, most innovative, and most successful concepts were incorporated into Khufu's pyramid. Better foundations, building on bedrock, the use of corbelled ceilings, and burial chambers built into the interior of the pyramid rather than in the bedrock beneath it were some of the advances later incorporated into the pyramids at Giza, the largest pyramids ever built. After

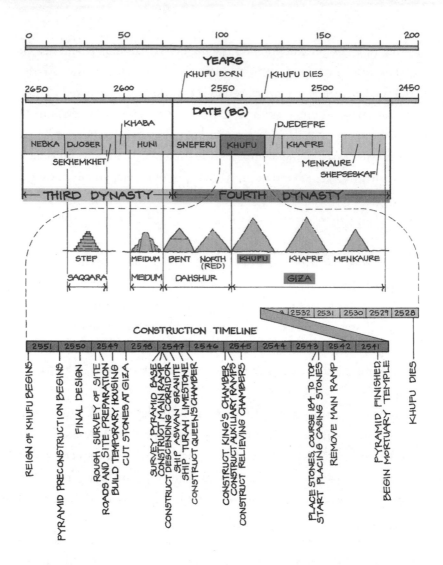

Fourth Dynasty chronology.

Giza, nothing was ever again attempted on such a scale. Giza was the apex of pyramid building. Lessons learned and skills developed for building massive stone structures were acquired during four generations of pharaohs in the Fourth Dynasty but then faded from the human record, lost in time. Seven pharaohs ruled during these eight decades, encompassing the Third Dynasty and the first half of the Fourth Dynasty. Their reigns left an indelible mark on civilization.

Dr. Hawass's permission affords us access to the Giza site well beyond public visiting hours. The time spent there in the early morning and evening is especially sublime. The wish I made during my first visit to this site has come true—we are alone with the ancients who created this spectacular place. What an extraordinary feeling of solemnity! As birds circle Khufu's pyramid at first dawn, as bats wing silently above it in the moonlight, I am absorbed by the beauty and solitude of this place—and am struck by the fact that this ground, sacred to the ancient Egyptians, has transcended the chasm of time.

I am awed once again by the magnificence of the builders' thinking and the bold and vast scale of their planning. What must it have been like to live among them and witness their creations? To live in a time when at least some significant segment of humanity could spend their days conceiving such splendor—not for themselves, but for a deity who was their leader here on earth? To know that by this extraordinary creation you were ensuring the smooth passage of your leader-god's spirit to the heavens? To know that for the son of the sun god you could guarantee a perfect transition to the afterlife?

Here, on the dusty slopes of Giza, this book is born. Go back forty-five centuries with me for the story of people whose lives were guided by an elaborate set of religious beliefs that formed the underpinnings of one of the greatest civilizations of all time—people who crafted their statement to history in spectacular form, creating the greatest structure on earth and celebrating the human mind and spirit in the process.

Join me as I reconstruct the stair steps to the gods.

EGYPT DURING THE OLD KINGDOM

The Great Pyramid at Giza! Anyone who experiences this magnificent structure up close cannot help being amazed—or at least enormously impressed—that an ancient race erected such an enduring, colossal wonder. For more than 4,000 years, it stood as the tallest structure ever built, its simplicity of form and precision of design and positioning imbuing it with an enduring power that has captured the imagination of humankind for centuries.

Constructed as the tomb of Pharaoh Khufu, the Great Pyramid stands as splendid testament to one who could conceive such a work—one who could inspire or command the dedication necessary to accomplish this monumental public works project. Henceforth I will refer to it as Khufu's pyramid, giving due honor to the man who created it as a fitting structure in which to be buried (See Plate 1.) In a general sense, the basics of how it was built appear obvious. Blocks of limestone were cut from a quarry nearby on the Giza Plateau and stacked up to create a towering structure. It is only upon considering this approach in detail that some of the attendant difficulties emerge and questions arise. How did the Egyptians, who had only primitive tools, cut and move huge blocks of stone? How many workers were required? How long did it take?

At the time of Khufu's reign, the population of Egypt was between 1 and 2 million.[1] This provides an upper limit on the human resources available to supply and feed a workforce for a huge public works project. While the population included artisans, laborers, craftsmen, and farmers, a large number of the able-bodied workers were required to feed both the pyramid builders and the rest of the population. However, the available workforce was equal to the task of building the pyramid.

It was also essential to have willing workers with the necessary skills. Having visited the tombs of the workers and artisans at Giza—those

Khufu's pyramid from the western cemetery.

whose own statements, written in their tombs, bespeak the pride they felt working on the pyramids—I find it inconceivable that slaves were involved. The tombs contain multiple generations. Father and son worked at Giza, and entire families were buried there. These people obviously took great pride in their work, and many of their tombs are smaller, pyramid-shaped versions of Khufu's tomb.

What could motivate people to dedicate their lives to such a demanding project voluntarily? I believe the explanation lies in the fact that construction of the pharaoh's pyramid was an act of national pride, a monumental achievement that symbolized the strength and power of Egypt. I liken it to the Apollo Space Program, undertaken by the United States under the direction of President Kennedy. In a way, the goals were similar: to undertake an enormous challenge, something that had never been done before—to reach out and touch the sky. As Jaromir Malek suggests, the large-scale building projects pushed by the pharaohs became a catalyst for change in Egyptian society.[2] And the fundamental forces that drove the execution of this extraordinary undertaking were rooted in the ancient Egyptians' culture and religion.

THE PREDYNASTIC PERIOD

Egypt occupies one of the most unique geographies on earth: the fertile valley of the Nile River. The annual flooding of the Nile brought a layer of rich black silt to the Nile Valley and Delta. At the same time, the flooding removed accumulated salts. Another quirk of geography—Egypt is bordered on the east and west by vast inhospitable deserts—combined with the fertility of the river, created ideal conditions for the emergence of a new civilization. In the distant past, wandering nomads found this fertile area and began to hunt and live there. The Paleolithic implements that have been found in Egypt show a gradual evolution paralleling that of Europe.[3] The archaeological record shows that stone tools became more and more refined.

The first "true" human, *Homo erectus*, lived several million years ago and is usually associated with the beginning of the Stone Age, or the Lower Paleolithic era.[4] *H. erectus* left Africa and arrived in the Middle East as early as 1.8 million years ago, probably transiting the Nile Valley on the river, a convenient route north. Some of these early humans no doubt settled in this fertile area.[5] *H. erectus* fabricated crude tools including scrapers and choppers. In the Middle Paleolithic era, H. erectus gave way to *Homo sapiens Neanderthalalensis*, also known as Neanderthal man. Some Middle Paleolithic sites have been found in Egypt in the Western Desert. There may have been sites in the Nile Valley, but they have been buried by the cycles of Nile flooding. The earliest known burial—that of a child—is thought to date to 55,000 years before the present time.[6] Then, about 40,000 years ago, *Homo sapiens sapiens* (often referred to as Cro-Magnon man after the site in France where remains were first discovered) emerged as the forerunner of modern humans. Most of what is known about *H. sapiens sapiens* comes from caves and burial sites in Europe;

these sites have lent their names to specific cultures—that is, Aurignacian, Solutrean, and Magdalenian. As these cultures evolved, tools and weapons became more advanced, and the first examples of jewelry and art appeared (15,000–8000 BC).[7]

It is interesting to consider the accomplishments of Cro-Magnon people in view of subsequent developments in Egypt. The Cro-Magnon improved tools and implements, including needles, fishhooks, and the bow and arrow. They developed communities that were based on a division of labor. They could count and possibly established a crude form of writing. They produced superb cave paintings. And they developed concepts of an afterlife, demonstrated by the care exercised in burying their dead and supplying their graves with jewelry, weapons, tools, and food.[8]

Many Late Paleolithic sites dating between 21,000 and 12,000 years ago have been found in Egypt. These include graves, remains of communities, hunting sites, and mines.[9]

Somewhere around 10,000 BC, the Paleolithic age gave way to the Neolithic period, and eventually stone tools came to be replaced by those made of copper. Other metalwork included copper vessels and ornaments of silver and gold. Neolithic people invented the arts of spinning, knitting, and weaving cloth. They made the first pottery and learned how to make fire. This age is remarkable because Neolithic people were the first to disperse themselves over the entire world, traveling immense distances by land and water to every corner of the globe.[10]

Early Neolithic people left many traces in the Western Desert of Egypt dating from 8800 to 4700 BC. These hunter-gatherers left evidence of temporary camps shared with domesticated cattle. Some of the camps had grain-storage pits and shelters, and in some cases they were permanent settlements. The camps also bore evidence of stone-grinding tools. It is now believed that after 4400 BC weather patterns underwent a dramatic change, and the arid climate made the desert less habitable.[11]

During the Neolithic age, several important human institutions first emerged: the family, the village, religion, and the concept of the state. Neolithic humans were the first to exercise some real control over their surroundings. Their food supply was more stable because they practiced agriculture and domesticated animals. They grew barley and raised sheep, goats, cattle, and pigs. They domesticated cats and dogs. They also crafted jewelry and painted pottery. By 4000 BC they knew how to smelt and work copper. They were excellent boatmen. No doubt these early inhabitants of the Nile Valley began their first efforts to deploy irrigation and to measure and understand the flooding of the Nile, a cycle upon which their lives depended. During this time the Neolithic inhabitants of the Nile Valley existed through a combination of hunting and agriculture; settlements up and down the Nile came at a later date. These people were not all of the same temperament. In the north, where the Mediterranean influence held sway, they were easygoing. In the south, under the influence of a harsher African lifestyle, they were driven and bellicose.[12]

It is not known for certain where the early Egyptians originated. The population may have been a mixture of immigrants and nomads who came to Egypt from the south (Nubia and Ethiopia), the west (Libya), and the northeast (Palestine and Syria), and who then merged and gave rise to the Egyptian people.[13]

The Stone Age came to an end at different times in different parts of the world. Portions of the New World—inaccessible jungle areas in Brazil, for example—were still populated by Stone Age people in 1950. In these remote spots, life went on much as it had 10,000 years earlier. We do not know for certain why some isolated areas never advanced. People with plentiful food and resources may have had more time to devote to arts and sciences not directly tied to subsistence. Whatever the reason, around 5000 BC the Neolithic era ended in the Nile Valley, and a more advanced civilization emerged. The Predynastic period, from around 5000 to 3000 BC, set the stage for the creation of some of the most awe-inspiring structures ever built by human beings.

The Predynastic period gave rise to two distinct cultures, the Badarian and the Naqada. The Badarian culture, which arose in Upper Egypt, boasted the earliest examples of agriculture, permanent settlements, and graves dating from around 5000 BC. Excavations from this period have unearthed pottery, tools, jewelry, statues, and artifacts of hammered copper.[14] During the Naqada period (4000–3200 BC), a ruler known as the Scorpion King is believed to have united several of the city-states in southern Egypt. The earliest examples of hieroglyphic writing date from this time.[15] The Scorpion King's reign was followed by that of King Menes (3000 BC), who is believed to have conquered Upper and Lower Egypt, uniting the country and setting the stage for the emergence of Egypt's first dynastic rulers.

The Naqada-period dead were buried in simple pits in the ground. Artifacts from this time include decorated pottery, carved mace heads, examples of holes drilled in hard stone, and the first examples of the ability to work granite, limestone, alabaster, and other stone materials. Copper working, no longer limited to small objects, came to be substituted for such stone tools as axes, blades, rods, and spatulas. Large quantities of copper ore found at Maadi may have come from Wadi Arabah, in the southeast corner of the Sinai Desert.[16]

By 5000 BC small villages were scattered up and down the Nile in Upper Egypt, and by 4000 BC they had been extended into Lower Egypt and the delta region. The Egyptian people were established along the Nile in a series of city-states called "nomes," each with a local ruler, or "nomarch." It is possible that the nomes originated near natural basins that occur along the length of the Nile.[17] These basins filled with water when the river flooded and retained the water until well into the dry season.

Many factors favored the development of the nomes and the eventual rise of Egyptian civilization. Foremost among them was the richness of the Nile Valley and Delta, where annual floods created a strip of exceedingly fertile soil along the more than 1,000-kilometer length of the river. The climate did not suffer extremes in temperature, and the river provided a broad and effective highway for commerce and communication. Boats used the river current to travel from south to north, yet for most of the year the prevailing wind is from the north, enabling travel upriver. The valley's inhabitants worshiped local gods and were engaged in agriculture, mining, manufacturing, and trade with other parts of Egypt and distant lands. Inhospitable deserts bordered the river channel on both sides, affording a degree of isolation and protection to the early Egyptians.

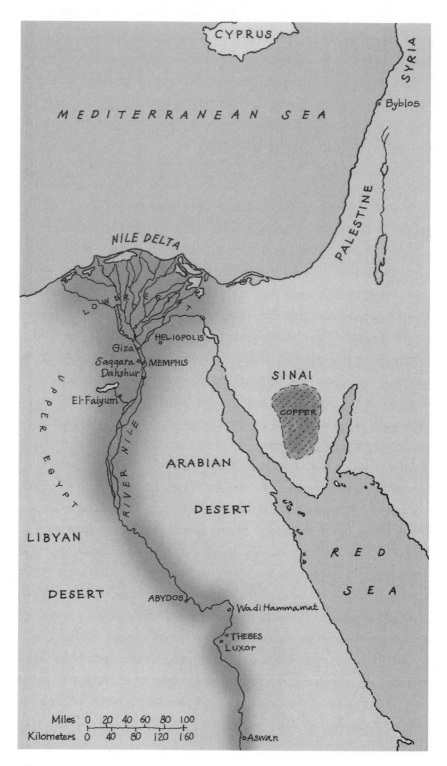

Upper and Lower Egypt and trade areas.

During the Predynastic period—perhaps for mutual defense against invaders—the individual nomes organized themselves into two kingdoms: one in the south (Upper Egypt) and one in the north (Lower Egypt). The Upper Kingdom extended from the ancient town of Elephantine, at the first cataract of the Nile, north to the Fayyum Basin; the Lower Kingdom extended from Memphis to the delta.

The center of activity in Upper Egypt was around Abydos or Thebes, where a local chief gained ascendancy over the southern nomes. Recent excavations near Abydos have located the graves of three of the kings who ruled Upper Egypt prior to the unification.[18] Around 3000 BC, another local ruler rose to prominence in Lower Egypt, in the area near Memphis. The period of progress for which we today recognize ancient Egypt began when the two kingdoms of Upper and Lower Egypt were unified around 3000 BC. This period became known as the First Dynasty and was followed by thirty more dynasties, until the time of the Greeks.

THE RISE OF THE PHARAOHS (3000 BC)

The unification of the Upper and Lower Kingdoms is credited to Menes, the first pharaoh of the First Dynasty. Menes brought together the "two lands" of Egypt, established a uniform set of laws, and extended communications and justice throughout the region. Henceforth the pharaoh was known as the "king of the two Egypts." The role of the pharaoh was to establish and maintain order and protect Egypt from its enemies. In a sense, the Egyptian world was isolated. It stretched for 1,000 kilometers along the fertile Nile Valley but was bordered on each side by barren deserts and at the extreme north by the unknown of the sea. Menes' capital was initially in the south, but at some point he recognized the strategic importance of being closer to the delta region and established a new capital at Memphis. Memphis retained its importance for centuries, even until the Fourth Dynasty. The pharaoh's residence was called the White Wall, a name that became synonymous with the capitol.

During this early dynastic period (First to Third Dynasties), nearly 400 years passed and more than a dozen known kings oversaw Egypt's development. As the country grew and prospered, trading expeditions were sent to Africa, the Middle East, and western Asia, seeking wood, precious stones, gold, metals, and livestock.

The pharaohs ruled Egypt for 3,000 years, beginning around 3000 BC and continuing until the time of Christ, when Egypt was conquered by the Greeks. Prior to the emergence of the pharaohs—or the period before the commencement of the dynasties—the Egyptians already had an advanced civilization. During the period encompassing dynasties four through seven, known as the Old Kingdom, Egypt would rise to even greater heights as the pharaohs grew in power.

When the pharaonic system became well established, the pharaoh ruled by divine right. Ancient texts describe the pharaoh as semidivine and imply that he enjoyed the favors of the gods and was protected by them.[19] By the Third Dynasty (the pharaoh Djoser and the Step Pyramid), the distinction between the king as the defender and organizer of the two Egypts, versus the king as a god, begins to blur.[20] In the funerary complex at Saqqara, the enclosure incorporated north and south palaces to represent Upper and Lower Egypt, and wall carvings show the king engaged in activities throughout the complex, symbolizing the role he played in maintaining order. This evolution continues with the pharaoh Sneferu at Dahshur. Finally, at Giza, the funerary complex clearly celebrates the pharaoh as a divine figure—now part of the sun god, the creator and omniscient ruler of the known universe. In the Fourth Dynasty, statues and carving represent the pharaoh on the throne with the falcon—Horus—symbol of the sun god, perched with him. The funerary complex enabled the pharaoh to remain alive on earth after death—as necessary, given his immortal status—and communicate with the gods in the afterlife.

During the time of the pharaohs, Egypt made many outstanding advances in the arts, law, mathematics, science, and religion. The Egyptians perfected tools, developed systems of irrigation, formulated a written language, and developed a solar calendar of 365 days. They observed the behavior of the star Sirius and noticed that once a year Sirius rose just before dawn—on a date coinciding with the annual flooding of the Nile. They fixed this time as the beginning of the calendar year, which was divided into twelve thirty-day months, and five days were added at the end of the year. Next, by observing the heavens, they divided night and day into twelve equal segments each. A segment was one-twelfth of the time between sunrise and sunset (or vice versa), so the length of the hour varied. Time was measured by a water clock—a bowl with a small hole in it—calibrated for the different seasons.[21] They also established a system of laws. The legal tradition that emerged was held in such esteem that Egyptian law was considered binding even on the pharaoh.[22]

The Nile River was a dominant force in the daily life of the Egyptians, who divided the year into three seasons according to the river's cycles. The time of flooding—the Inundation—typically ran from July through October. During this time farmers and peasants worked on such civil projects as improving dikes or canals or building public works, including the pyramids. As soon as the waters began to recede in September, farmers entered the fields for a light plowing and planted seeds in the mud. Livestock were driven into the fields to churn the soil. Crops of emmer wheat, barley, lentils, and beans were planted. The second season—the Emergence of Fields from Water—ran from November through February. At this time water was carefully managed: the river basins were drained, and water was stored in catch basins and ponds. The soil remained moist until February. Irrigation ditches brought additional water to crops. The third season—the Drought—ran from March through June, the time for harvesting and threshing.[23] An overarching system of taxation lent strength and stability to the Egyptian state. Royal surveyors measured a farmer's field to determine how much of his yield would be turned over to the king. The royal granaries provided a hedge against bad times. If the crops failed in one region for some reason, the pharaoh could send grain to resolve the crisis. Large agricultural surpluses were possible. With the storage and distribution of grain controlled by the state, Egyptian civilization was able to flourish.[24] The annual cycle was important even for tax purposes, because land was divided into three categories: always flooded, sometimes flooded, or never flooded. Flooded land bore the highest rate.

The pharaonic government was a well-organized system. The pharaoh was the high priest; the priests were his chief subordinates. He was also served by a vizier, or prime minister, tax collectors, a treasurer, and a system of governors for local districts who reported to him and oversaw the process of tax collection. Forty-two nomarchs, or local governors, twenty-two in Upper Egypt and twenty in Lower Egypt, were appointed to ensure that the pharaoh's wishes were carried out. The administrative system also included officials who served as judges in regional courts.[25]

The vizier reported to the pharaoh and in some cases held multiple titles, such as overseer of all the king's works; seal bearer of the king; overseer of fields, gardens, cattle, peasant farmers, and granaries; and steward of the king. Broadly stated, his duties could be divided into three categories: he was the king's deputy, the head of civil administration, and the managing director of the royal palace. The vizier presided over the highest court and served as the chief justice. He saw to the collection of taxes from all over the land and, in later periods, the collection of tribute from abroad. He was responsible for security and maintaining order. In his daily routine, he would have met with the treasurer to review finances and with other key personnel—the keepers of the granaries and foremen of major construction projects, for example.[26]

Also important to the smooth running of the kingdom were the numerous scribes employed by the king, the temples, or wealthy nobility. Scribes had to serve a lengthy apprenticeship, practice writing sayings and stories, and learn the spellings of everyday items. They might survey a farmer's field to levy taxes or accompany a caravan to a foreign country as part of a trade mission—to the Sinai for copper or precious stones, or to Beirut or Byblos for lumber.[27] And they could practice another profession, such as law, taxation, engineering, or architecture.

The crown prince was educated by scribes and priests for his future role. He was required to master hieroglyphics, the intricacies of religion, and the significance of the multiple deities recognized at that time. He served under his father in some significant position, in charge of public works or one of the state's projects in mining or agriculture. It fell to the crown prince to complete his father's works upon the pharaoh's death.

As early as the First and Second Dynasties, sea trade with Lebanon intensified. Inscriptions have been found that record the visit of Khasekhemwy, the last pharaoh of the Second Dynasty, to the ancient city of Byblos.[28] Mines were operated in the Eastern Desert, producing gold, copper, and semiprecious stones. Quarries up and down the length of the Nile produced a variety of stone for building construction and the decorative arts, including alabaster and high-quality white limestone at Turah in the south, basalt at Memphis, and granite and diorite at Aswan. With a unified, centralized government, increasing wealth, the rise of skilled artisans and builders, an emerging sophistication brought about by contacts with the outside world, and blossoming capabilities in mathematics, astronomy, and measurements, Egypt was fully ready to undertake impressive public works projects.

The first elaborate underground tombs began evolving in the middle of the First Dynasty. These tombs were the forerunners of the Step Pyramid. From the First and Second Dynasty tombs found at Abydos and Saqqara, we note the inception of features later found in a royal funerary complex: chambers, boats, jewelry—all the supplies necessary to support the pharaoh in the afterlife. Tombs imitated a royal residence, and later in the period a temple appeared, possibly a precursor to the mortuary temples found in the Fourth Dynasty. Still, these First and Second Dynasty tombs were modest compared to those built later and lacked any substantial superstructure or building on top of the burial chamber.

RELIGION

Religion, a means of explaining the unknowable, played an important role in Egyptian life and helped shape the pyramids.[29] Egyptian religion is in itself a complex subject. Most of what we know about it comes from later dynasties, because information about religion during Khufu's time is limited.

The Egyptian gods were associated with the creation, the forces of nature, and the health and welfare of the people. Celebrations were held and offerings were made to ensure successful crops and good health. Religion was also of great importance because it held out the promise of an afterlife.

From the earliest recorded times, the Egyptians believed in a single, all-powerful, immortal god, Neter, who had created the earth and all who inhabited it. The Egyptian religion endured for thousands of years, and while it always maintained this concept, over time the central belief was overlaid with a belief in other, lesser deities. Yet even as these polytheistic beliefs came and went, the monotheistic core remained.[30]

Any detailed understanding of Egyptian religion is complicated by the fact that scholars have drawn on bits and pieces of information written by different authors in different ages. Nowhere is there a single, comprehensive statement of religious thought. This, combined with the fact that during its 3,000-year history the religion of the ancient Egyptians continuously evolved and became more complex, makes a detailed understanding very difficult. It is as if we tried to understand Christianity, with its numerous variations (Catholicism, Protestantism, etc.), by looking at a few grave inscriptions, a hymnal, a Baptist tract, and so on without having the Bible as a reference.

The Egyptian myth of creation held that before the earth was formed, there was only Nun, the god of primeval water. The earth is portrayed as a mound that rose from Nun, and from this mound came Atum, the "old" sun god, later called Ra, who generated himself from Nun. Ra's offspring were Shu, the god of the air, and Tefnut, the goddess of moisture. Shu and Tefnut in turn gave birth to Geb, the earth god, and Nut, the sky goddess. The offspring of Geb and Nut were Osiris, Isis, Seth, and Nephthys. (See Appendix 1 for a listing of the principal gods.)

The "pyramid texts"—Old Kingdom records inscribed in tombs and pyramids—and other documents recount the legend of Osiris, who ruled the entire world as a benevolent king but was tricked and cruelly murdered by his brother Seth. Seth mutilated the body and scattered the pieces, but Isis found the pieces and used her magic to restore Osiris to life, and thus he became king of the region of the dead. A variation of this legend holds that Ra ordered the body embalmed, and then Isis restored it to life. She conceived a son, Horus, on her husband's mummified body. Horus later killed his uncle and avenged his father's death.

There are various legends concerning Ra's travels through the sky that bear upon the design of pyramids. Ra was thought each day to emerge or be born in the east and to set or die in the west. One idea was that the god flew through the sky on birdlike wings. Another was that the sun was carried across the sky by a giant beetle, just as a scarab rolls a ball of dung. The most interesting concept suggested that the sky was formed by the body of the goddess Nut, who arched over the entire earth, her head in the west and her groin on the eastern horizon. Each evening she swallowed the sun, and during the night it moved through her body, to be reborn in the morning in the east.[31] (See Plate 10.)

In the New Kingdom, other gods were brought into the religion, causing it to become even more complex. Consistent throughout Egyptian religion, however, is the concept of the all-powerful sun god, Ra. Ra was the symbol of God, the supreme deity. Next in importance to Ra was Osiris, the god of resurrection. Osiris suffered death and mutilation but rose again to become the god of the underworld and the judge of the dead. The belief in Osiris was critical to the Egyptian treatment of death and the afterlife. By Osiris's example the righteous could look forward to an afterlife in paradise—as long as the body of the deceased were properly prepared and provisions were made for sustaining it in the Netherworld. Likewise, the importance of Ra, who was reborn each day and who brought about Osiris's resurrection, can be understood in the context of pyramid construction.

The pharaoh was considered the sun god's representative on earth and served as an intermediary between the people and the god. Later, perhaps beginning in the Third Dynasty, the distinction between intermediary and god changed. The pharaoh then became divine, the sun god on earth.

Many interesting accounts of preparing and safekeeping the body of the deceased have been discovered. The pyramid texts, for example, include rituals, hymns, and magical texts.[32] Following the Eighteenth Dynasty, collections of magic spells and rituals were written on papyrus and left in the tombs.[33] *The Book of the Dead*—a compilation of spells and magic formulas written on papyrus or tomb walls to protect and aid the deceased in the afterlife—describes how to avoid corruption of the body and ensure its progression into the afterlife, when judgment, referred to as the weighing of the heart, would take place. *The Book of the Dead* replaced the earlier pyramid texts but contained many of the same spells found in the earlier collections.

Egyptian religion embraced the concept of a divine judgment after death.[34] The judgment scene, which is depicted with slight differences in various papyri, shows the heart of the deceased being weighed on a balance in the hall of Maat, the goddess of truth. The heart is placed on one pan of a balance scale and weighed against a feather, which is Maat's sacred emblem of truth and goodness. The dog-headed god, Anubis, adjusts the balance. Horus stands nearby, ready to conduct the deceased into the presence of Osiris, who will judge the decedent's worthiness to enter paradise. Thoth, a scribe of the gods, frequently represented as ibis-headed, stands by to record the decision. "The eater of the dead"—a god with the head of a crocodile, the forebody of a lion, and the hind quarters of a hippopotamus—stands ready to devour the poor sinner if the heart is found to be heavy. If the heart is true, the individual is allowed to enter paradise. Heaven had two parts: the Field of Rushes and the Field of Offerings—counterparts to the two lands of Egypt.[35] (See Plate 14.)

In the Egyptian concept of the afterlife, the deceased was expected to eat, drink, and live a life of pleasure in paradise, with a body not unlike the one he or she inhabited on earth.[36] From this, it is logical to see how the custom of mummification, or preserving the earthly body form, grew in importance. A dismembered, burned, or incomplete body was unsuited to success in the afterlife. The form of burial was influenced by the Egyptians' concept of the body, which consisted of several parts, each of which had to be cared for in the afterlife. The *khat*, or physical remains (corpse), had to be preserved from decay or destruction to protect the *ka*, or spiritual "double" (personality or life force) of the individual, believed to exist separately from the khat. The *ka* could move from place to place and enter heaven to speak with the gods. It required sustenance, so it was important that the tomb, or pyramid in this case, be supplied with a means for the *ka* to go in and out and receive food and drink. The deceased's family or estate made arrangements for provisions to be brought to the tomb or temple periodically for this purpose. The *ba*, or the individual's soul, dwelt in the *ka* but was another part of the total person. It was believed to have the power to leave the tomb, assume different forms (it is frequently pictured as a human-headed hawk), and enter heaven to live in a state of glory with the gods. The *khu* represented the spirit of a man and, like the *ka*, could be imprisoned in the tomb, so special measures were designed to prevent this. Some measures were largely symbolic, such as the false doors that were carved as part of the tomb to tell the *khu* that provisions had been made for its freedom. The *sekhem* was a person's power or strength. The *khaibit*, his shadow, was associated with his *ba*. Finally, the *ren* was a person's name, an important part of the larger being of a person.[37]

This overview of Egyptian religion is of necessity brief and incomplete. It was a religion that continued to evolve, and a full description would depend on the date and place under consideration. My purpose here is to provide a basic understanding of the religious elements that would have driven certain aspects of the design of the Great Pyramid. As the spiritual and temporal leader of the Egyptian nation, the pharaoh underwent detailed training in religious matters, so these elements would have been familiar to Khufu.

Another tradition that characterized Egyptian religion was the construction of tombs to protect the remains of the deceased and the belongings deemed essential in the afterlife. This was as true for the burial of common people, who might be buried in humble graves in the desert sand with a few meager belongings, as it was for the pharaohs. The lavishness of these final arrangements increased in cost and complexity in accordance with the wealth and importance of the deceased. With mummification, the Egyptians developed elaborate means of protecting the body of the deceased on its journey to the afterlife. As these ceremonies evolved in complexity, so did the tombs and funerary complexes prepared to support the rituals. Thus the tomb became more than just the final resting place of the deceased: it was also the means by which the spirit of the dead was sustained over time with offerings of food and ceremonies celebrating the life and accomplishments of the individual, and it protected the remains and spirit of the deceased.

Early tombs featured a flat bench or platform—a *mastaba*—on top of the grave. These tombs evolved as more lasting forms of burial mounds. It is believed that the earliest pyramids evolved from mastabas and reflected this custom on a larger and much grander scale, fitting for the king. *Mastaba*, the Arabic word for "bench," has been used to describe this type of construction. (See Plate 2.) Later, additional levels or "steps" were added to the mastaba. The stepped construction reached its apex with the Step Pyramid of Djoser at Saqqara, which rose to a height of 60 meters and could have symbolized a stair step to the gods for this Third Dynasty ruler.[38] We next encounter the early Fourth Dynasty pyramids constructed by Sneferu at Dahshur, which built upon Djoser's work and include the Bent Pyramid and the North—or Red—Pyramid. The latter is the first true pyramid constructed on a large scale and certainly served as the inspiration for Khufu's pyramid. After all, Sneferu was Khufu's father.

The evolution of tomb design from mastaba to pyramid.

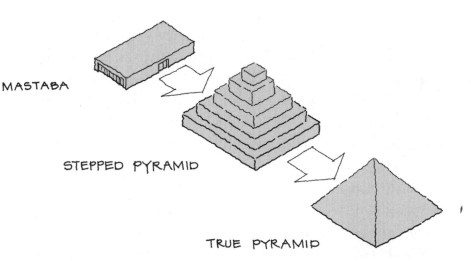

MASTABA

STEPPED PYRAMID

TRUE PYRAMID

How can the evolution from the stepped-pyramid construction to the true pyramid be explained? There is probably no way of knowing for certain. The true pyramid's precisely oriented and sloped internal corridors offered a passageway for the spirit of the pharaoh to rise and communicate with his fellow gods in the heavens. Given the fact that the pharaoh came to be considered a divine being, the literal stair steps of the stepped pyramid become superfluous. The true pyramid, with its smooth sides, can be said to more perfectly represent the rays of the sun streaming to earth to envelop the tomb of the sun god on earth, the pharaoh. Additionally, it presents a daunting challenge to the world, defying imitation or duplication.

Viewed within the context of the ancient Egyptians' religious beliefs, the pyramid's various components are now more easily understood. Mutilation and decay of the corpse were to be avoided so that the *ba* could function. The preparation of the corpse by mummification satisfied part of this requirement; entombment in the pyramid, where the *khat* was protected against grave robbers and others who might be tempted to disturb the carefully preserved remains, was another. The tomb—be it within a pyramid or elsewhere—also served as the place to store supplies: the food, personal items, gifts, and magical spells needed by the *ka* of the deceased in the afterlife.[39] Passages were incorporated into the tomb so the *ba* could escape and travel into the sky to become part of the celestial world. In the pharaoh's case, the passage was constructed to direct his *ba* to the northern circumpolar stars, where it could join the other gods in the heavens. False doors were built into the tomb to provide an exit to the east, in the direction of the sunrise. For the wealthy, the tomb contained "living quarters," chambers where the deceased could live in the afterlife as on earth. It was incumbent upon relatives to deliver food and other necessities to sustain the deceased in the afterlife.

At Giza, the pyramid complex also included a mortuary temple, where priests and other officials could symbolically attend to the needs of the deceased king. The mortuary temple was a symbol of the pharaoh's everlasting power as a god-king.[40] The mortuary temple was connected to a valley temple by a long causeway. The valley temple had access to the Nile by a canal. Only traces remain of the mortuary and valley temples associated with Khufu's pyramid, but Khafre's mortuary and valley temples at Giza are well preserved and provide solid insight into what these structures were like.

The Old Kingdom (2600–2100 BC)

The Egyptian economic system was largely agrarian. The peasants were free and able to own land and livestock. Most worked for nobles and large landholders, and they had their own small plots as well.[41] Crops included emmer wheat, lentils, barley, flax, sesame, castor beans, date palms, and papyrus; livestock included cattle, sheep, swine, and goats.[42] Manufacturing industries were established to produce ceramics, cloth, tools, and jewelry. They also made bread, wine, and beer. As early as the beginning of the Old Kingdom, large numbers of people were involved in the manufacture of crafts and the production of goods, including glass, pottery, textiles, and ornamental objects. There were also skilled stonemasons and shipwrights. Workers worked shifts of ten-day weeks, followed by two days off. Wages were paid in bread, beer, beans, onions, dried meat, fat, and salt. The wage scale was based on skill level. For example, foremen and scribes were the highest paid, followed by draftsmen, sculptors, and painters, then quarrymen and stonemasons, and at the bottom of the scale, the unskilled laborers who dug or hauled materials.[43]

By 2600 BC Egyptian trading vessels bearing cargoes of lentils, grain, textiles, papyrus paper, and other products ventured into the Red Sea and along the Mediterranean shore to trade. Overland expeditions traveled south to Nubia.[44] Papyrus became an important agricultural product. In addition to being made into paper, it was used to make baskets, mats, sieves, and other products. The fibers could be twisted into cordage and used to make rope. In fact, Egyptian rope was used on many of the vessels that worked Mediterranean waters. Egypt had a monopoly on the production of paper until the modern form was invented.[45]

Much of the economy was based on barter or exchange of goods, yet early on the Egyptians developed such business instruments as accounting systems, tax records, contracts, deeds for property, and wills. They developed a system for written records on papyrus paper. Later they introduced a metal standard for large transactions. Taxes were collected kingdomwide to support the royal establishment and public works, and tribute was levied on neighboring nations.[46]

Marriage was common, and family activities occupied an important part of daily life, as seen from the family groupings in tombs of both commoners and royalty. Sons frequently followed in their father's footsteps. Generations of painters, sculptors, or metalworkers are buried in workers' tombs at Giza. Women had rights, owned property, and were held in high regard. And these people enjoyed life: their art reflects parties, celebrations, singers, dancers, musicians—and abundant food and drink. Each morning the priests paid homage to the local gods in the temple, the same ritual followed by the pharaoh on behalf of the entire kingdom. The priests often had multiple duties. In addition to their religious activities, they were scribes, physicians, teachers, or surveyors. The year was marked by festivals, of which perhaps the most important and universal was the harvest festival. Others were dedicated to local gods, and plays were performed in honor of Osiris. In the Festival of the Valley (at Thebes), the pharaoh crossed the Nile in the royal boat to pay homage to his ancestors buried in the western hills. During the festival of Amon, images of the god and his wife, Mut, were carried by water from Karnak to Luxor and back again.[47]

The Old Kingdom was characterized by a well-developed judicial system. Regional judges heard cases and made sure that the laws laid down by the pharaoh were applied uniformly to royalty and the common man. There was, however, no large-scale standing army in the Old Kingdom. Because of its isolation, Egypt was not threatened by invaders. The military was limited to small units tasked to guard supply routes, escort caravans and trade expeditions to mines in the Sinai and Wadi Hamammat for metals and precious stones, and supervise work gangs. It is believed that the nomes had local militias but that much of their activity was dedicated to public works. In times of national crisis, the pharaoh could call upon the nomarchs to provide military forces to operate under a national commander. In general Egypt existed peacefully, in part because of its isolated and protected position and in part because the union of Upper and Lower Egypt had been the result of mutual benefit and cooperation, rather than aggression and conquest.[48]

Engineering, mathematics, and science—disciplines necessary to perform large construction works—were clearly applied in ancient Egypt. Egyptians made early advances in astronomy and mathematics toward practical and religious ends rather than pure science. Their accomplishments included the development of a calendar based on astronomical observations. They identified major stars and could identify the position of the stellar bodies with some accuracy. In mathematics they were able to calculate areas and volumes of complex objects and such complex structures as the pyramids. They understood design and the laws of scaling (see drawing on page 76). Artists used scaling and grid lines when painting scenes. To make a grid line, a cord coated with red ocher pigment was snapped against the surface to be painted. The same method was used in the quarries to lay out blocks of stone to be cut. Blue, red, yellow, green, black, and white pigments were ground to powder and mixed with water, beeswax, albumin, gum, or some other medium and applied as tempera on plaster after it had set. For black, lampblack was used. Ochers (clay soil bearing traces of iron ore) made brown, red, and yellow. Powdered malachite (copper ore) made green; chalk or gypsum, white. For fine work, a brush made from a thin reed was used. For coarser work, brushes were made from palm sticks with frayed ends or bundles of grass that were tied together.[49]

Measurements were carefully made using a cubit rod (similar to a yardstick, but 52.4 centimeters long). The cubit was divided into seven *palms*, each four *digits* long. Thus, there were 28 digits per cubit.[50] The ancient Egyptians used many basic tools made of copper, including saws, chisels, hammers, and drills. They understood the principles of the lever, the inclined ramp, and the use of rollers, but wheels were not used in the Old Kingdom.

The most common building materials were mud bricks, which the ancient Egyptians were known to use prior to 3000 BC. In fact, the Egyptian word for brick was *dobe*. Thousands of years later, the Spanish adopted this term as *adobar* (to plaster), which led to the term familiar to us: *adobe*. The Egyptians dug a dark gray Nile alluvium and mixed it with sand or chopped straw as a binder to make bricks. The bricks were cast in wooden molds, and the brick makers "struck" the brick (knocked it out of the mold) onto a flat surface to dry in the hot sun. The Egyptians knew about firing bricks but generally used sun-dried bricks in construction because they could be made more easily and quickly and did not require scarce fuel. I have seen mastabas and walls at Giza, Dahshur, and Saqqara made from mud bricks that are still in good shape after more than 4,000 years, so the brick makers' judgment was sound. Bricks were made in a standard size for thousands of years: approximately 20 digits long by 10 digits wide by 6 digits thick (36 by 18 by 11 centimeters). Through experience gained with mud bricks, Egyptian masons learned methods for laying and bonding masonry and for constructing arches and barrel vaults.

Mud bricks were widely used for houses and other buildings. Panels could be constructed with reeds or thin sticks as reinforcement. Roofs were spanned with timbers or poles. A layer of matting was laid down, and the surface was covered with thatch or a layer of mud. Ceilings in finer homes or estates were made from another layer of matting that was plastered and then painted. Floors were dried mud covered with a hard layer of gypsum. Windows were small and placed high in the walls. There was very little furniture in the typical house.[51]

Good native wood was scarce in ancient Egypt, so carpenters became skilled at producing usable boards by assembling smaller pieces from the available local woods: sycamore, fig, acacia, tamarisk, sidder, and willow. Wood was fabricated into boats, furniture, and utensils. Joining was done using straight or tapered hardwood pegs. Egyptian carpenters knew common joining techniques, including overlapped miter joints, lap joints, mortise-and-tenon joints, dovetail joints, rabbet joints, and dadoes. Cutting and boring tools were made of copper: the saw, adze, axe, chisel, knife, scraper, and bow drill. Quartzite was used as a whetstone to sharpen tools. Measuring instruments—squares, levels, and cubit rods—were made from hardwood. A lump of sandstone served as a plane. Fine sand was used for finishing. Glue, wood filler, and a type of varnish were known.

As early as the Second Dynasty, wood was imported when required in large quantities or in better quality—for instance, for use in construction projects. A "timber fleet" operated between the delta and the Syrian coast, bringing in cedar, cypress, fir, and pine from Lebanon. Timber was also rafted down from Byblos. Sudanese ebony was brought down the Nile. Good wood was reused.[52] The Nile was the principal means of transporting goods. During the year it provided easy access over the length of the kingdom. Vessels or barges coming north traveled down on the current or were rowed if necessary. Vessels traveling upriver against the current were sailed (the prevailing wind conveniently blew from the north) or rowed. During the period of inundation, water spread over a great area, and shallow draft boats or barges could reach remote areas, including the pyramid construction sites.

The ancient Egyptians were skilled sailors and boatbuilders. Limestone reliefs at Saqqara dating from 2650 BC show various aspects of boatbuilding. Vessels ranged from small punts, canoes, or rafts made from reeds lashed together to mammoth barges capable of carrying upwards of 100 metric tons of stone. Freighters carried grain up and down the river. The hulls of the larger vessels were made of timber and were carvel-built (flush planking) with a low, curved prow and a high stern. There was a central mast, a lateen sail, a steersman in the stern with a steering oar, rowers amidships, and perhaps an awning for shade. For a large vessel, there might be forty to fifty rowers; for a small one, between two and five. A splendid example of such a vessel is in the boat museum at Giza. It was discovered in one of the boat pits next to Khufu's pyramid and was removed and carefully reassembled. Its overall length is 43.3 meters, the beam is 5.9 meters, and the draft is 1.48 meters. Ancient records report a raft 30 meters long by 15 meters wide. With a draft of 1 meter, such a vessel would displace 450 metric tons. A vessel 45 meters long was used to transport obelisks. Sneferu had a vessel 100 cubits long.[53]

On land, transportation was by foot or donkey. The nobility traveled by carrying chair or palanquin. The carrying chair recovered from the tomb of Queen Hetepheres, Khufu's mother, is a beautiful example.[54] Horses and chariots came much later and were not used during the Old Kingdom.

Thus it is reasonable to assume that the Fourth Dynasty Egyptians possessed the skills and resources necessary to design, plan, and conduct a project as complex as the Great Pyramid at Giza. The planning and design would have included the preparation of models or drawings, calculating the volume of materials needed, developing detailed schedules and budgets, and anticipating the logistics of such a complex activity, which would undoubtedly involve the labor of thousands of people over a period of several years.

THE PHARAOH BEHIND THE GREAT PYRAMID

During the thirty to forty years that preceded Khufu's reign, the stage was set for the construction of the Great Pyramid. The questions remain, however: Who could have imagined a project of the scope and magnitude of the Great Pyramid at Giza? Who could have mandated the resources necessary to make it happen? Who could have had the vision and the

Pharaohs of the Third and Fourth Dynasties

THIRD DYNASTY	2649–2575 BC	FOURTH DYNASTY	2575–2465 BC
Nebka	2649–2630	Sneferu	2575–2551
Djoser	2630–2611	Khufu	2551–2528
Sekhemkhet	2611–2603	Djedefre	2528–2520
Khaba	2603–2599	Khafre	2520–2494
Huni	2599–2575	Menkaure	2490–2472
		Shepseskaf	2472–2467

The Fifth Dynasty begins with the pharaoh Userkaf, in 2465 BC. There is a two-year gap in the ancient records. We do not know if Shepseskaf had a short-term successor or whether Userkaf's reign was just delayed.

Source: Lehner (1997), 8–9.

FOURTH DYNASTY FAMILY TREE

		SNEFERU 2575–2551 1st pharaoh, 4th Dynasty	+	HETEPHERES Daughter of King Huni 3rd Dynasty			

ITET (or ATET)	PRINCE NEFERMAAT Vizier to Sneferu	KHUFU 2551–2528 2nd pharaoh 4th Dynasty	HENUTSEN Half-sister to Khufu	Several older brothers	NEFERTIABET Khufu's sister

(Vizier) →

HEMIUNU
Vizier to Khufu
Grandson of Sneferu

(Vizier) →

KHUFU & MERITITES	KHUFU & ?	KHUFU & HENUTSEN

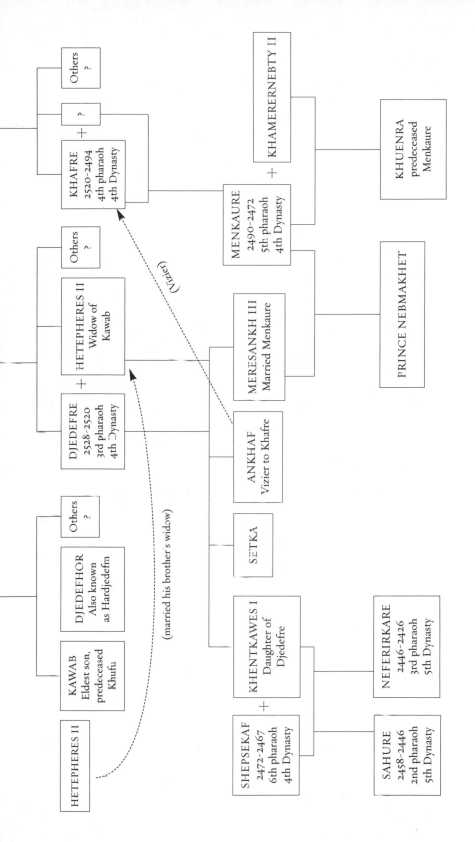

staying power to see it through to completion? To answer these questions, we need to take a step back in time, to the Third Dynasty.

In the Third Dynasty, Djoser ruled as the second pharaoh. Under his direction and with the guidance and planning of his grand vizier, the famous Imhotep, the Step Pyramid was built at Saqqara in an effort to make a mastaba grand enough for a pharaoh. We can surmise that having seen the stepped pyramid take shape, Imhotep, or perhaps the pharaoh himself, visualized smooth surfaces on the side of the pyramid, the apex rising to a peak.

Perhaps models were built; perhaps smaller versions were constructed and then demolished to test the concept. No definitive evidence has yet been found, although I. E. S. Edwards describes two pyramid models—one of the Step Pyramid, the other, a conventional pyramid—carved from limestone.[55] What we do know, however, is that following the construction of the earlier pyramids (the Step Pyramid and the Meidum Pyramid), the Bent Pyramid was constructed, and during its construction the design was altered to reflect what was being learned by trial and error. Given Imhotep's reputation, it is conceivable that he led, or at least encouraged, further innovation. Thirty-six years elapsed from the end of Djoser's reign to the onset of Sneferu's reign, so Imhotep may still have been alive for part of this time. (It is thought that he died during the reign of the pharaoh Huni.)

The lessons were duly noted and recorded, and during the reign of Sneferu, the first pharaoh of the Fourth Dynasty, further improvements in design and construction evolved. Perhaps this came from the influence of Imhotep, or possibly from one of his protégés. Nefermaat, one of Sneferu's sons, may have learned his skills from Imhotep or one of his subordinates, or at least felt his influence, but for whatever reason, Sneferu made Prince Nefermaat his vizier.[56] Another of Seneferu's sons, Khufu, became the second pharaoh of the Fourth Dynasty. Then Prince Nefermaat's son Hemiunu (who was also Khufu's cousin) became Khufu's vizier. As Khufu's vizier, Hemiunu saw many important works constructed, but clearly his greatest was the Great Pyramid. It is also noteworthy that the builders of the second and third pyramids at Giza were Khufu's son Khafre and his grandson Menkaure. A family tree shows the relationship between a number of these Fourth Dynasty rulers and their wives, viziers, and heirs.[57]

Sneferu was the most prolific pyramid builder, for he constructed not one but three major pyramids. His first was an eight-step pyramid at Meidum modeled after Djoser's Step Pyramid. In the middle of his reign, Sneferu moved to Dahshur, and there he built two large pyramids, the Bent Pyramid and the Red (or North) Pyramid. Near the end of his reign, he had his workmen return to Meidum, and there he initiated work to convert the Meidum pyramid from a stepped pyramid to a true pyramid. Whether the work was ever completed we cannot tell for sure, because layers of stone were stolen from the pyramid over the ages, and the outer layers collapsed, leaving the stepped core structure we see today.

Three of Sneferu's sons are buried at Dahshur. His fourth son, Khufu, inherited the throne. Having watched his father's works take form, he no doubt had a vision of what he wanted to achieve when he began construction of his own pyramid. I find it significant that he did not continue building at Dahshur. This may have occurred for a number of reasons. First, the problems with the Bent Pyramid demonstrated the importance of selecting a site with good underlying bedrock. Additionally, it is reasonable to believe that as a new king Khufu wanted something that labeled his reign as unique, something that would distinguish his from his father's long, successful reign. The two large pyramids built by his father at Dahshur stood 200 cubits high. Why build in the shadow of these if he could select a new site, with ample stone for construction nearby, and build a pyramid 250 or even 300 cubits high? Confronting this challenge, his ally, his confidant, and the person entrusted with oversight of the project would certainly have been his vizier.

One wonders about the decision to proceed with the construction of Khufu's pyramid. It must have involved delicate and diplomatic discussions, since the subject must have reminded the pharaoh of his own mortality. Yet reality demanded that to maintain the power of the king and his succession, his immortality must be prepared for in the traditional manner. Perhaps there was even an appeal to Khufu's ego.

We can imagine Pharaoh Khufu thinking about the project, knowing that it was something that he must do, and resolving that it would solidify his position as pharaoh. After all, he was young for the position. A project of significant scope and magnitude would enhance his position with the people and occupy the time and attention of the royal court, leaving less time for intrigue or palace gossip. But whom would he entrust with such an important task?

There could be only one choice—Hemiunu, vizier, overseer of all the king's works. Dependable, solid, trustworthy, a man who attended to details. Khufu's cousin, absolutely reliable, would get the job done. Hemiunu would form his program management team, breath life into the project, inspire his aides with its scope and grandeur, and use his royal authority when necessary to draft or conscript the other labor, skilled workers, or resources needed for this vast undertaking. Once

under way, the project would gather momentum and take on a life of its own. It embodied the national prestige of Egypt. Foreign visitors would be impressed, if not cowed, by its immensity. Once it was under construction, all of Egypt would willingly participate, anxious to help in some way. For any who were less than willing, Hemiunu no doubt could bring to bear the high authority of his office. With knowledge gained through decades of prior building experience, the emergence of an energetic new king, and the selection of his vizier to oversee the project, the stage was set to build the Great Pyramid of Khufu.

The pyramids at Giza viewed from the southeast.

CHAPTER 2

PYRAMID EVOLUTION

The construction of Khufu's pyramid was not just the next logical step in the evolution of the increasingly bold and sophisticated stone-construction technology that distinguished pyramid architecture, it was also the pinnacle of this evolution, which achieved its first major landmark with the construction of the Step Pyramid at Saqqara. By trial and error, Third and early Fourth Dynasty builders gradually perfected the design concepts and construction techniques that made Khufu's pyramid possible. The skills and techniques were passed down from one generation to the next over a period of eighty years—from the start of Djoser's reign to the reigns of Sneferu and Khufu—just as the power to rule Egypt was passed from one generation to the next. Eighty years is a sufficiently short span of time for a construction supervisor on one project to be alive for the next and possibly a third—or at least to pass on the skills and the lessons learned to the next generation of builders so that there would be some continuity in expertise.

The most significant pyramids of the Third and Fourth Dynasties illustrate the evolution of construction technology during that 184-year period (see the drawing opposite). Each is imposing in its own way, and each is subtly different. Clearly the builders tried new ideas, and as they learned from their mistakes and gained confidence, they pushed their tools, the materials at their disposal, and their engineering and construction techniques to ever more impressive accomplishments.

The man who planned Khufu's pyramid certainly studied these earlier efforts, seeking to discover what worked and what pitfalls to avoid. After all, his objective, like that of his predecessors, was to push the design and construction envelope further—literally to heights never before achieved. Khufu's pyramid would become a symbol for the vitality and power of the new king's reign. Travelers from afar would look at it

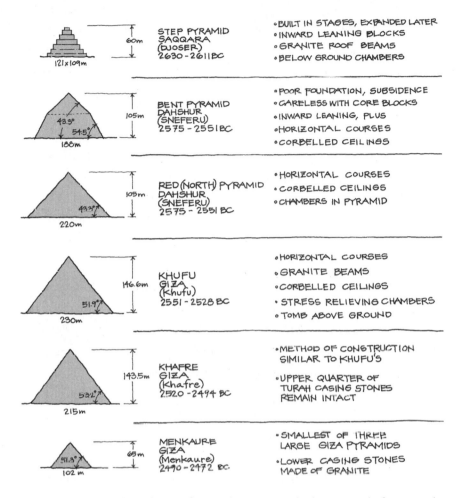

STEP PYRAMID SAQQARA (DJOSER) 2630-2611 BC	60m 121x109m	• BUILT IN STAGES, EXPANDED LATER • INWARD LEANING BLOCKS • GRANITE ROOF BEAMS • BELOW GROUND CHAMBERS	
BENT PYRAMID DAHSHUR (SNEFERU) 2575-2551 BC	105m 43.3° 54.5° 188m	• POOR FOUNDATION, SUBSIDENCE • CARELESS WITH CORE BLOCKS • INWARD LEANING, PLUS • HORIZONTAL COURSES • CORBELLED CEILINGS	
RED (NORTH) PYRAMID DAHSHUR (SNEFERU) 2575-2551 BC	105m 43.3° 220m	• HORIZONTAL COURSES • CORBELLED CEILINGS • CHAMBERS IN PYRAMID	
KHUFU GIZA (Khufu) 2551-2528 BC	146.6m 51.9° 230m	• HORIZONTAL COURSES • GRANITE BEAMS • CORBELLED CEILINGS • STRESS RELIEVING CHAMBERS • TOMB ABOVE GROUND	
KHAFRE GIZA (Khafre) 2520-2494 BC	143.5m 53.2° 215m	• METHOD OF CONSTRUCTION SIMILAR TO KHUFU'S • UPPER QUARTER OF TURAH CASING STONES REMAIN INTACT	
MENKAURE GIZA (Menkaure) 2490-2472 BC	65m 51.3° 102m	• SMALLEST OF THREE LARGE GIZA PYRAMIDS • LOWER CASING STONES MADE OF GRANITE	

The evolution of pyramid construction leading up to Khufu's pyramid.

in wonder and know the grandeur that was Egypt and the power that was Khufu.

Hemiunu was no doubt aware of past project failures—the collapse at Meidum, the settlement and cracking at the Bent Pyramid. It is logical to assume that he made efforts to learn from the past. Perhaps he himself traveled to Dahshur and Saqqara, examining the existing pyramids, looking at them from a new perspective: how could they be improved upon? At the very least, he would have summoned scribes, architects, builders, historians, and priests who served in Sneferu's temples as well as other elders who were familiar with the history of these places. He would have asked them to reveal all they knew about the great monuments that stood at Dahshur and Saqqara: what worked, what failed, what should be done differently.

THE STEP PYRAMID (2620 BC)

The Step Pyramid was a Third Dynasty pyramid constructed under the direction of Imhotep, the vizier and chief architect of the pharaoh Djoser. No records survive to show what it looked like in finished form. It was robbed of most of its limestone centuries ago, even before Napoleon's arrival in 1798. The sketches published in 1809 in Napoleon's magnificent chronicle, *History of the Expedition*, show the Step Pyramid as it appears today.[1] The Step Pyramid is significant because it reflects advances in and techniques for building stone structures to greater heights than previously attained. The approach taken in the initial plan was to build a mastaba, later converted to a five-step pyramid, and then to its final form as a six-step pyramid. Rather than being placed horizontally, the inner blocks slope inward. When the additional step was added, the size of the pyramid was greatly increased. It is the pyramid with by far the greatest expanse of underground tunnels and chambers, a truly remarkable example of underground work. The underground chambers have flat ceilings; Djoser's tomb has granite roof beams. Access to the burial chamber was through a cylindrical opening, which was plugged with a round granite cylinder weighing many tons.[2] All of the underground chambers are in the soil beneath the pyramid. (See Plate 4.)

The Step Pyramid is in a large walled enclosure or courtyard, some of which has been restored. There are other tombs and structures within the enclosure, and on the southeast corner the remains of a causeway angle off toward the Nile River. In an enclosure on the north side of the pyramid, a replicated statue of Djoser peers out through a peephole.

The masonry and stonework of the Step Pyramid are of a poorer quality than that used in Khufu's pyramid, and the blocks are smaller. They are loosely stacked in inwardly leaning layers and backfilled with

The southeast corner of the Step Pyramid, showing how the pyramid was expanded.

sand, dirt, and clay mortar. On the south face are remains of casing stones (around 15 to 20 centimeters thick) that are smoothly polished and trimmed. They vary in length from 38 to 76 centimeters and are about 20 centimeters high. Some are embedded in the wall, so you can see where the layers were added when the pyramid was enlarged. The core blocks are irregularly shaped stones, not cut smoothly, perhaps 56 centimeters long by 20 centimeters high by 40 ± 5 centimeters deep.

On the southeast corner, one can clearly see the joint where the original Step Pyramid ended and additional layers were added. Additionally, three pieces of wood resembling logs protrude about 1.2 meters from the side of the pyramid. It is the only example of wood remaining in the construction that I observed in any of the pyramids I examined. The purpose of the beams is not known, but I surmise they were used during construction.

There are many other smaller pyramids at Saqqara, as well as numerous mastabas and tombs constructed in the Fifth and Sixth Dynasties. A short distance to the northeast of the Step Pyramid is the mastaba of Ti, a Fifth Dynasty overseer of pyramids and temples. This mastaba consists of an open courtyard with a descending corridor that leads underground to a series of superbly decorated rooms and chambers. It is one of the most beautiful tombs I have visited. The walls are decorated with bas-relief carved and painted panels that depict the deceased, his servants, and scenes from everyday life. One panel shows in detail how boats were constructed; another how bread was made; still another, how pottery was produced.[3]

THE PYRAMID AT MEIDUM (2570 BC)

Next in the evolution of pyramid design came the pyramid at Meidum. The height of this pyramid increases to 92 meters at the traditional 52° angle. Originally a pyramid of seven steps, it was constructed by Pharaoh Sneferu. Later in his reign, the exterior surface of the pyramid was converted to a true pyramid shape. However, the inner structure still consisted of blocks leaning inward. (See Plate 5.)

The pyramid at Meidum represents another technological advance: It is the first pyramid to incorporate corbelled ceilings in the underground chamber. The corbelled ceiling design accommodates very elegant arched ceilings that are suited to the strength of the limestone building material. This design eliminated the need for heavy and much harder-to-cut granite beams and allowed for longer ceiling spans.

Today, when viewed from a distance, the pyramid at Meidum resembles a central masonry tower of three steps protruding from a pile of rubble. It was in this form when visited by Napoleon's expedition, so the casing stones—the fine white exterior limestone—must have been stolen centuries ago. There is speculation that the outer layers of the pyramid collapsed at some point in time. Whether this happened during the Old Kingdom or later, as thieves attempted to remove the casing stones, is not known.

The central core (the portion still standing) must have been skillfully constructed, since these interior casing stones are still intact. Nicely finished casing stones can be seen in places, placed with tight joints and of better quality than the exterior casing stones, which were added later. The stones above are quite large and stacked in a random pattern. They are not precisely cut, vary in shape, and are arranged in an imprecise geometry. On the northwest corner, where the outer layers have crumbled down and disintegrated, it is still possible to see the pyramid's approximate shape before the erosion took place. I measured one of the exterior casing stones where the rubble had been removed and the stone was partially uncovered. It had been carefully trimmed to an angle of 48° from the horizontal.

At the corner I could see the extension of the walls for the casing stones on the west and north, even though the corner is broken. And up inside—higher—perhaps 15 to 20 meters back, casing stones delineated another, smaller version of the pyramid. The lower casing stones were horizontal—roughly 1 meter wide and about 38 centimeters thick. Up higher, they appeared to lean inward.

It was on a Friday afternoon, the last day in March 2000, when I entered the pyramid. The opening is partway up on the north face, where a lot of erosion has occurred. The initial portion of the Descending Corridor is lined with finely trimmed stones, forming a rectangular tunnel about 1.6 meters high and a little less than a meter wide. Farther down, the tunnel is rough-cut and not trimmed, its curved ceiling made of blocks of stone. The excavation was made, then lined with stone. The upper portion is in the body of the pyramid, while the lower portion is at or slightly below grade, beneath the base of the pyramid.

At the bottom of the Descending Corridor is a horizontal passageway, then a vertical shaft rises into a burial chamber with a corbelled ceiling. There, the quiet and coolness is disturbed by workmen, who, in the

The causeway at the Meidum pyramid.

glare of powerful lamps, are drilling holes in the ancient masonry to strengthen the ladder that provides access to the chamber. The burial chamber is small, the corbelled ceiling executed more crudely than the work we see in later pyramids.

The other unique feature of the Meidum Pyramid is the very clearly defined remains of a long causeway, cut into the bedrock at the level of the pyramid enclosure and sloping easterly down in the direction of the Nile River. During construction it was used as a road to transport barged materials to the job site. Once the pyramid was finished, it provided access during ceremonies.

The Bent Pyramid at Dahshur (2565 BC)

Following the construction of the Meidum Pyramid, Sneferu and his architects began constructing an even grander pyramid at Dahshur. This pyramid began with a base of 188 meters and a slope of 54.5°. The Dahshur site consists of sandy desert soil, a poor foundation for the pyramid without adequate bearing capacity for the massive structure. This pyramid also began with inward-leaning blocks. As construction progressed, subsidence was observed and the load produced by the inward-leaning blocks caused structural problems within the pyramid. Inward-leaning blocks, originally adopted to provide stability in the mastaba design, were found to contribute to excessive loads in the core of the structure. Consequently, the architects and builders reduced the slope of the pyramid by approximately 10° one-third of the way up and continued construction with an angle of 43°. This resulted in the "bent" appearance of the pyramid. Corbelled ceilings were also employed in the Bent Pyramid. Another important innovation was that the new core blocks (at the levels above the bend) were laid horizontally as opposed to leaning inward. (See Plate 6.)

On the east face of the Bent Pyramid, near the northeast corner, the surface is still finished and smooth. Some of the casing stones remain in place. I noted a number of small, random marks going up the face, each of which has been carefully plugged with a piece of matching stone. Could they have been made simply to correct defects in the stone, or

The Bent Pyramid.
Note the sandy soil at the base.

were they something more complicated—perhaps holes to support scaffolding or some rigging used by the stonemasons as they put the finishing touches on the smooth surface? Since there doesn't seem to be any pattern in their placement, any uniformity in location or distance, scaffolding does not seem likely.

Within the Bent Pyramid is a burial chamber with a corbelled ceiling reached by the Descending Corridor, entered on the north face. This chamber was placed below the pyramid in the desert soil. The Bent Pyramid differs from the other pyramids discussed thus far in that it has a second entrance on the west side that leads to another burial chamber above the first chamber. According to Mark Lehner, this chamber was threatened by the settling that took place and was shored up with heavy wooden beams.[4] This early attempt at building a burial chamber in the body of the pyramid ended in failure.

The Red (North) Pyramid (2560 BC)

Sneferu apparently was not satisfied with the Bent Pyramid, although it is 105 meters high and was the most striking structure built at that point in the pharaonic era. He proceeded to construct a new pyramid north of the Bent Pyramid at the same site. According to marks found on some of the stones, it was constructed "in the thirtieth year of his realm."[5] The Red (North) Pyramid was designed with a slope of 43.3° and a height of 105 meters, equivalent to the height and the top slope of the Bent Pyramid. However, in this case the designers were more careful with their choice of a site and obtained better foundation and bearing conditions for the pyramid. In addition they used horizontal placement of core stones and corbelled ceilings for the interior chambers. (See Plate 7.)

On the western face it is possible to see a corner block of limestone

Casing stones, Red Pyramid.

that bears surveyor's marks lining up the base mat for construction—that is, setting a corner angle (see photo on page 140). Also, the intact casing stones are very closely fit and have been carefully finished with a smooth surface. Thicker and more uniform than the casing stones of the Bent Pyramid, they are placed horizontally. The reddish-pink limestone from which the name Red Pyramid is derived was mined from a quarry not far to the southwest of the pyramid. Faint traces of a supply road that once led from the quarry to the pyramid are still visible. On the east face, a portion of the casing is intact clear down to the platform level. Some of the casing stones bear cutouts and notches, some filled in, some not. Levers were inserted in them to position and maneuver the stones into place. The backing stones behind the casing are irregular in shape and size, and the spaces are filled with mud mortar or chinked with odd pieces of stone to level up the course.

On the east side of the pyramid is a small pedestal constructed from blocks of stone. Mounted on top of it is the reconstructed capstone or pyramidion of the Red Pyramid. It is approximately 1 meter square at the base.

The entrance to the Descending Corridor of the Red Pyramid is positioned high up on the north side. I paused there before beginning the descent down the long, sloping corridor. From this vantage point I could see the surrounding desert. Today it is dry, barren. I wondered what it looked like forty-five centuries ago as preparations were being made to entomb Sneferu's body. Partway down the corridor, the air was suddenly cooler. I was reminded of a line from a Carl Sandburg poem: "in the dust, in the cool tombs."[6] Descending farther, I entered the first of the three chambers. Each had a corbelled ceiling. From the first antechamber, I passed directly into the second. Even in the dim light I

could see that the chambers were splendid early examples of this corbelling technique. From the second chamber, I climbed a ladder to another short tunnel and through that entered the third chamber, believed to
be the actual burial chamber. It is above and turned at a right angle to the
two antechambers, and it is larger. This chamber had the most impressive
example of a corbelled ceiling I'd seen so far. (See Plate 8.)

In the Red Pyramid, for the first time, the builders succeed in locat-
ing the king's burial chamber above ground in the body of the pyramid.
In all earlier pyramids, the king's burial chambers were far enough below
ground that they would not be damaged by the weight of the structure
standing above, with the exception of the failed attempt in the Bent
Pyramid. In the Red Pyramid, the designers had enough confidence in
their structural design to place the King's Chamber above ground.

The experience gained in building this pyramid for Khufu's father
was brought to bear in the design and construction of Khufu's pyramid,
for which it was the prototype. The lessons learned from the Red
Pyramid that influenced the design and construction of Khufu's pyramid
several decades later included:

· Site. It must be solid rock: no shifting sands, no possibility of
settlement.

· Foundation. It must be strong enough to support the weight of
the structure above and not be crushed or dislodged.

· Measurements. They must be very precise to maintain the shape of
the true pyramid, plumb and true to a height never before achieved.

· Burial chamber. It should not be beneath the pyramid in the cold
rock but in the heights of the pyramid itself so that the pharaoh's *ba* can
freely communicate with the gods and he will forever experience the
warmth of Ra.

· Masonry courses. They must be placed horizontally to better
distribute the loads and avoid creating stresses in the interior chambers
and corridors.

· Structural stability. Massive blocks must be used at the lowest levels
to create an accurate and stable base, but it is not necessary to use them
in every course, because smaller blocks will be easier to manage and

place. Periodically courses with large blocks should be repeated to over-lap and structurally tie together the construction.

· Ceilings. They will all be of corbelled construction, since that gives the most spacious and dramatic effect while providing the structural strength required.

· Inclination (*seqed*). To minimize labor and materials and maximize height, it should be as steep as possible, but not steep enough to create structural or construction problems. (The Red Pyramid was successful with a *seqed* of 7 palms, 2 digits, while the Bent Pyramid encountered problems at a *seqed* of 5 palms.) But with a better site and foundation, it should be possible to improve upon the Red Pyramid, so the work will be planned for a *seqed* of between 5 palms, 2 digits, and 6 palms. (See Appendix 2 for a definition of *seqed* and its conversion to angles meas-ured in degrees.)

· Casing stones. The pyramid will be finished with the finest white limestone, so it will be visible from a great distance.

We can be certain that Hemiunu's designers and architects were aware of these concepts and considered them for inclusion in the design of Khufu's pyramid.

Sixty years passed from the end of Djoser's reign in 2611 BC to the onset of Khufu's reign in 2551 BC. Therefore it is possible that living rep-resentatives transmitted knowledge and experience gained in the earlier construction projects from one reign to the next. If not directly trans-mitted by individuals, the information was possibly handed down from father to son, or from master to apprentice. We do not know if Djoser's chief architect, Imhotep, was still alive at the time Khufu's pyramid was built. (It is believed he died during the reign of the pharaoh Huni.) However, it is certain that foremen, overseers, architects, and other skilled personnel from the reign of Sneferu were still living when Khufu began his construction. The unique nature of the Red Pyramid would have ensured its use as a model for future monolithic construction, and those who were involved in its construction would have been sought out for their opinions on how to build more efficiently and avoid problems experienced in the past.

Khufu's Pyramid at Giza (2540 BC)

Records indicate that Khufu began his reign in 2551 BC, following the death of his father, Sneferu. Sometime after assuming the throne, Khufu ordered the construction of a new pyramid, this time on the Giza Plateau. Khufu's pyramid became the grandest accomplishment of the Fourth Dynasty and certainly is the most magnificent representation of the art of stonemasonry and pyramid building produced by the ancient Egyptians (see photo on page 29). Most significantly, Khufu's pyramid reflected the culmination of the learning experiences and trial-and-error adjustments made by the previous pyramid builders. It featured horizontal construction of its core masonry, the backing stones, and casing stones. It rose to an unprecedented height of 146.6 meters with sides that sloped 51.9°. Khufu's pyramid contains what is perhaps the finest example of a corbelled ceiling to be found in ancient Egyptian architecture. (See Plate 11.)

Placing the foundation on the bedrock of the Giza Plateau was an improvement in that it provided a firm base that would eliminate settling and other structural problems. While much of the site was carefully leveled, a portion of the bedrock was left as a protrusion into the body of the pyramid. This offered several advantages. It reduced the amount of tedious labor required to level the site. It eliminated the need for quarrying, transporting, and placing a number of the huge blocks that formed the lower courses of the pyramid. (The bedrock was cut in stair step fashion to match the size of the stones placed next to it.) Finally, this protrusion of bedrock served as a huge key to lock the structure in place at the site and give it added stability against any form of ground motion.

The northeast corner of Khufu's pyramid, showing platform stones and the bedrock massif.

Granite beams were employed in several of the interior chambers. Another innovation found in Khufu's pyramid is the stress-relieving chambers above the king's burial chamber, which accommodated the tremendous weight of the pyramid above the king's final resting place.

THE OTHER PYRAMIDS AT GIZA (2540–2480 BC)

Khufu's sons carried on his legacy at the Giza site. Records found with the wooden boats entombed at Khufu's pyramid indicate that his son Djedefre oversaw his father's funeral and then succeeded him as pharaoh. Djedefre reigned only eight years and was replaced by Khafre,

Khafre's pyramid at Giza. Note the casing stones on the top one-third.

Menkaure's pyramid at Giza.

another son of Khufu. Khafre was responsible for the second major pyramid at Giza, which rose to a height of 143.5 meters. It did not quite reach the height of his father's pyramid but is itself another outstanding example of pyramid construction during the Fourth Dynasty. Khafre's legacy is the best-preserved funerary complex and mortuary temple at the Giza site, as well as the enormous sculpture that has attracted worldwide attention for centuries: the Sphinx.

The third major pyramid at Giza was constructed by Khafre's son Menkaure. Although Menkaure's pyramid reached a height of only 65 meters, it is an extraordinary example of pyramid construction because of the red granite casing stones that were used on the lower courses. It is a pyramid in which many of the casing stones have survived, and thus we can infer details on how the construction was performed and how the finished angles were achieved with such accuracy. (See Plate 3.)

Following the reign of Menkaure, the era of constructing immense pyramids came to an end. Never again in the subsequent dynasties would structures of such scale and grandeur as those at Giza be built. While many other tombs, temples, and lesser pyramids were built over the course of the next several thousand years, the most extraordinary period of construction in ancient Egypt had ended, and the quality of stonemasonry gradually declined from that point on.

The execution of such massive and complex projects at Giza required an intricate series of preparatory steps. Naturally, it would be impossible to simply show up at an inhospitable site in the barren Western Desert with an army of workers and start building. First, the project needed a leader. Hemiunu met the requirements for this demanding role. Next, the project demanded a great deal of planning, a thorough analysis of the work requirements, the completion of initial measurements and surveying, the refinement of the design, the acquisition of the needed tools and materials, and the preparation of the site—first to support a large workforce, and then to ready it for actual construction. (See Plate 12.)

PLANNING

The Egyptians clearly possessed a sophisticated ability to plan complex projects; it is otherwise inconceivable that a construction project of this scale could have been undertaken in the middle of the desert. Limited evidence indicating the planning ability of ancient Egyptians survives to this day. While no written planning documents survive from the Fourth Dynasty, there is ample indirect proof of their ability to plan and organize complex tasks.

The administration of the Egyptian kingdom itself by the time of the Fourth Dynasty was a complex undertaking. A communication system obviously existed by which the pharaoh could receive news from outlying provinces and send instructions to his agents and representatives. Taxes were levied and collected, a practice implying the existence of a system of record keeping. A well-developed calendar was in use, so activities could be scheduled for auspicious or appropriate times. Egyptians understood the timing of the annual flooding of the Nile and made use of this information for planting and harvesting crops. Finally, the builders could draw on knowledge gained during the construction of earlier pyramids, including the Step Pyramid and the Bent Pyramid, in planning a large, complex public works project.

ANALYSIS

The planning of such a complex project requires the analysis of many different parameters. The amount of material required was an initial consideration. Before a site could be selected, it was essential to determine that the required building materials could be obtained within a reasonable distance from the site. An estimate of the size of the labor force was also required to determine that the required workers, both skilled and unskilled, could be found, housed, and fed. Myriad other details needed to be considered, such as water supply, how to bring ships close to the site, and the area required for the construction of ramps.

Egyptian mathematics were advanced, and we know that the ancient Egyptians could calculate areas, angles, and volumes. In fact, records of such calculations have survived from later periods. One example is the *Rhind Papyrus*, which describes a series of arithmetic and geometric calculations.[7] The Egyptians were able to carry out the accurate dimensioning and layout of large structures. They could calculate the weight of large objects and structures and determine the number of blocks required for a sloping embankment.[8] They used a decimal system and had established symbols for 1, 10, and 100 cubits, for example. They had developed methods for multiplying and dividing. Division was accomplished by breaking quotients into a series of sums of fractions—accurate enough for practical purposes. The concepts of squaring and finding a square root were known to them. They had an approximate method of determining the area of a circle, which was to subtract one-ninth of the diameter and square the remainder: $A = (8D/9)^2$. This gives the equivalent value of π (pi) as 3.16, rather than 3.14. They knew how to calculate the volume of cylinders and pyramids. Among the more sophisticated calculations the ancient Egyptians made was finding the volume of a truncated pyramid, obviously of great importance in determining the volume of material needed and the labor required, among other practical matters.[9]

MEASUREMENTS

Egyptians developed measuring instruments as early as the First Dynasty. The basic instrument was the cubit rod. A cubit was defined as the length of the pharaoh's arm from his elbow to the tip of his fingers. During the Fourth Dynasty at Giza, the cubit was approximately 52.4 centimeters. Various cubit rods have been found, dating from different dynasties, and their length varies slightly, so it would appear that there was no absolute standard.

The cubit is subdivided into 7 palms, each with 4 digits. A digit, defined as the width of the pharaoh's finger, is about 1.88 centimeters.

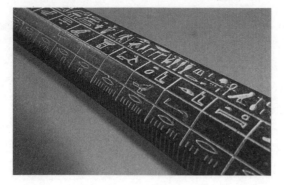

A replica of a cubit rod.

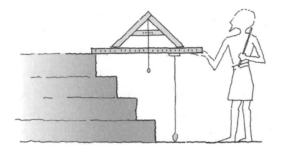

Measuring instruments, including various types of levels.

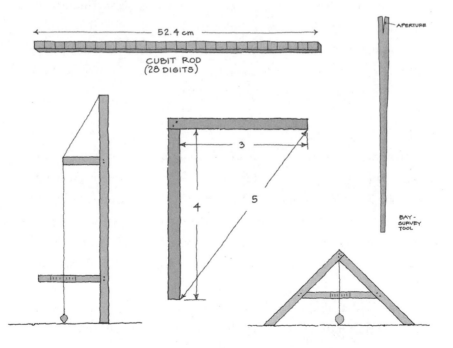

52.4 cm

CUBIT ROD
(28 DIGITS)

APERTURE

BAY-
SURVEY
TOOL

3

4

5

The digit was further subdivided into eighths, tenths, and so on to provide a very accurate smaller scale. For surveying and quarry work, longer rods of 3 to perhaps 10 cubits in length were used. For still greater distances, a system of "rods and chains" was used, although the chain was actually a woven rope calibrated with knots or marks and cut to a precise length, probably 1 *khet* (100 cubits). One common unit of area was 100 by 100 cubits, known as a *stat*. Systems for measuring angles were known, since the slope of the pyramid was controlled very accurately, as was the rise and decline of the ascending and descending tunnels inside the pyramid. Also, these were no random angles, because they were reproduced quite accurately in other works.

Measurement tools included the square level, used for leveling horizontal surfaces. It was shaped like the letter A with a plumb bob hanging down from the apex of the A. When the surface was level, the plumb bob was directly in the center of the cross bar on the A. A plumb bob could also be used to create a vertical level. We also know that the Egyptians had a square, or a means of making a right angle, most likely developed from the principle of a 3:4:5 triangle. A triangle constructed with its sides in these proportions will contain a 90° angle.

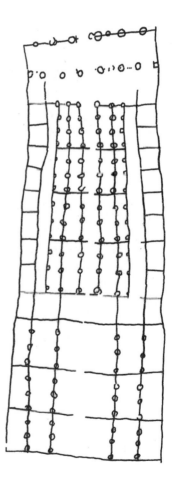

DESIGN

No records of designs survive from the Fourth Dynasty, but various plans and sketches have survived from later periods. For example, Dieter Arnold has compiled a list of more than a dozen,[10] including the floor plan of a temple from the Amarna period. In other examples,

The floor plan of a temple.

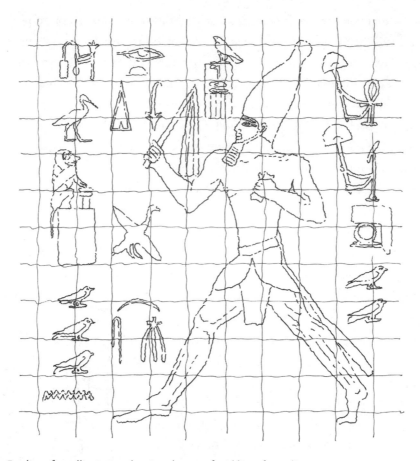

Replica of a wall painting showing the use of grid lines for scaling.

sketches or dimension lines indicate where cuts were to be made, which demonstrates that the concept of a written or drawn plan was known. Ancient designers made drawings that showed both elevation and plan views. Somers Clarke and R. Engelbach speculate that plans may have been preserved in palace archives. Models of structures have also been found.[11]

That the Egyptians understood the concept of scale drawings is evident from unfinished tomb paintings, in which the artists have constructed a square grid pattern to help establish the scale for the work and assist in laying out the figures. Construction lines cut into the limestone at Giza survive to this day, indicating the dimension of the blocks the stonemasons were to cut. Construction lines produced with red paint indicating horizontal alignment, leveling, and centerlines can be found on Menkaure's pyramid (see photo on page 184).

Thus, while no plans per se for Khufu's pyramid have ever been found, one must have existed, if only in the mind of a master builder, who would have conveyed it to the foremen and supervisors by means of sketches crudely drawn on scraps of stone or sketched in the sand at the site. Or a model may have been built. Certainly, verbal instructions were given in the field to the workers. While the pyramid contained a dozen or so special structures within a structure, the basic structure itself was highly repetitious. In other words, once the methodology for laying the first several courses was firmly established in the minds of the work crews and their overseers, it was simply a matter of stacking up more stone while carefully controlling the dimensions and the slope of the outer edges.

SURVEYING

Egyptian techniques for surveying were based on the use of a knotted cord for measuring length and the square level. It is also possible that wooden rods 3 cubits or longer were used. The ancient Egyptians were able to level a site with precision and could lay out the base of a large structure very accurately. There is good evidence of the high level of skill they possessed at Giza, where the dimensioning and leveling of the site came close to what could be achieved by modern methods.

For leveling, it has been suggested that the site was banked up and water placed in ditches to provide a reference plane.[12] Modern thought discounts this approach because of the size of the area, the dry climate, water loss through seepage into bedrock cracks, and the distance water would have to be hauled (not to mention the volume of water required). The Egyptians' lengthy experience in constructing irrigation systems establishes their skill in leveling. It may have been that they used this knowledge to level the site. Rather than use water-filled ditches, long wooden containers could have been employed. However, no example of such a device has been found.

It appears more likely that a s benchmarks was set along the sides of the pyramid base and a leve ce point established by sighting, using the square level described earlier. This line was then extended the distance required, and the foundation platform, or first course of blocks (rather than the bedrock), was leveled accordingly. The bedrock was probably cleared and roughly leveled, but the accurate leveling was done on the foundation platform, where small discrepancies could be compensated for easily. Lehner, who presents a thoughtful description of this process in his book, pointed out to me the line of rectangular holes that runs along the east and north sides of the pyramids.[13] These could have held posts to which a leveling line was attached and pulled taut (see photo on page 140).

The average length of each side at the base of the pyramid, based on the most recent data, is 230.4 meters (see drawing opposite).[14] The deviations from the average are as follows:

North side: - 10.97 centimeters

East side: + 3.35 centimeters

South side: + 8.84 centimeters

West side: - 0.61 centimeters

The greatest variation is between the north and south sides, where the difference is 20 centimeters. Thus, the errors range from roughly 1 part in 2,000 to 1 part in 38,000. On a level surface, it would be possible to achieve measurements of this precision with two accurate 10-cubit measuring rods laid end to end twenty-two times. Such a rod would be 5.2 meters long. Therefore, the accuracy (in the worst case of 11 centimeters of error) would require not misplacing the end point any more than about 5 millimeters for each of the twenty-two measurements, which seems possible. This is certainly within the accuracy of current rod and chain measurements, which can be made with an accuracy of 1 part in 100,000 with suitable care and corrections.

The right angles on the base of the pyramid deviate from a true right angle by up to 3.5′. (Two of the corners are very close to 90° exactly.) Clarke carried out an experiment by sighting with a large right-angle square and was able to set a right angle with a leg of 175 cubits with an

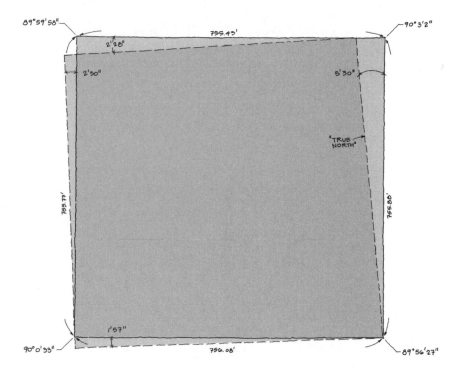

Base dimensions of Khufu's pyramid (in feet).
Note: The deviations from a true square are exaggerated for clarity.

accuracy of 1.5′. This suggests that the accuracy obtained was within the limits of Egyptian technology.[15] Modern surveying would achieve accuracy of ± 0.05′ to 0.5′.

It is interesting to note that the south side is almost exactly 440 cubits long, which suggests that this might have been the basic dimension planned for the base, since the Egyptians typically used whole-cubit measures in laying out large structures and even rooms within structures, if space and work conditions permitted.

In addition to achieving precision in leveling, alignment, and measurement, they were also skilled at accurate orientation. This is confirmed by the fact that the axes of the pyramids at Giza are not only closely aligned with each other, but also with the cardinal points of the compass: north–south and east–west. This could have been accomplished by celestial sightings (using a circumpolar star) or by sighting on the sun.

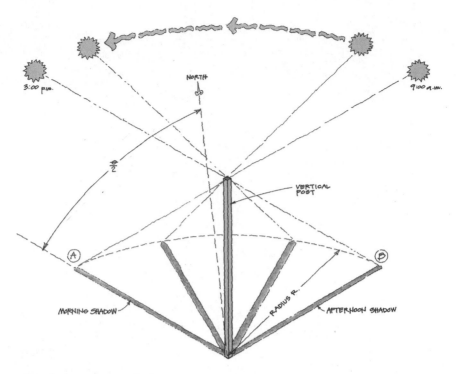

Sun-shadow survey method.

The star method is described in Edwards (1993) and numerous other works.[16] However, because the Egyptians worshiped the sun, it seems plausible that they preferred the sun method, which is easier to implement. Of two techniques using the sun, the simplest is the sun-shadow method, which only requires a plumbed (vertical) staff and a piece of string.[17] The second method requires knowing when the sun's altitude crosses a specific angle in the morning as it rises and again in the afternoon as it sets. This is easily done with a transit, but the surveyor could also sight the sun with a pointing device fixed at a predetermined angle, although at some risk of damaged vision!

With the sun-shadow method, the shadow of a tall, vertical pole is marked on the ground with small stakes every thirty minutes or so from 9:00 a.m. to 3:00 p.m. A smooth curve can be "drawn" through the marks with a piece of string to mark the path of the sun's shadow. Another piece of cord centered at the vertical pole is then used to draw a circular arc, which will be found to intersect the shadow curve at two points, A and B (see drawing opposite). A straight line can then be drawn from A to B. The midpoint of this line is due north of the vertical pole. Some error is present because the sun moves diagonally across the equator instead of following an exact east–west path. The pyramid is not oriented exactly north–south and east–west. The east side is 5′30″ west of north; the west side is 2′30″ west of north.

I performed a crude experiment with the sun-shadow method near my home at Balboa Beach, California, at latitude 33°36″ north, where the magnetic deviation is 14° east. (Giza is at latitude 30° north.) The method yielded true north with an accuracy estimated to be ± 1° (that is, 60′). In other words, the Egyptians were about twelve times more accurate.

It would be necessary to align only one side precisely (say the east side), and then the other side could be set at right angles to the baseline using the tools described above.

MATERIALS

In many ways the Fourth Dynasty represented the apex of Egyptian stonemasonry. While exquisitely delicate and beautiful structures and statuary were produced in later periods (for example, the temples at Karnak, 1300 BC), never again were such massive structures constructed. By the Fourth Dynasty, Egyptian stonemasons had perfected the art of cutting and finishing stone. Although they worked with several different materials, they preferred limestone for the massive pyramid structures because it was readily available and soft enough to work easily. By trial and error they came to understand the limitations of limestone in terms of its load-bearing capacity and strength, and therefore they knew when it was necessary to use a stronger material such as granite to support greater loads. They had learned that thick limestone beams, twice as deep as they were wide, for example, could span 2.5 to 3 meters, but for greater distances granite had to be used. They also had developed techniques for making mud bricks and gypsum mortar, although the mortar used at Giza does not appear to have been of a very high quality.[18] Mud brick walls as high as 13.7 meters and 24.4 meters thick were known.[19]

Wood was not available in large quantities, and special types of wood were imported from other places such as Lebanon. Skill at woodworking during the Fourth Dynasty is clearly evident from the boats found at Khufu's pyramid, as well as from other artifacts and tomb paintings.

The ancient Egyptians made rope from local vegetation; this is evident from the rigging on the boats and from tomb paintings. Ropes of various diameters were made, including some with a working strength of 5 tons.[20]

The availability of metals was limited. They primarily used copper and gold. Copper could be worked into tools but was generally soft. The Egyptians may have known how to temper or harden copper by heating and hammering, but no solid evidence of this practice has been found. Modern excavations at Giza have found remnants of copper and even the hearths or kilns used to process the metal into tools.

Egyptian experience with copper predates the pyramid era by at least a thousand years. Hammered copper objects were found in Predynastic times (4400–4000 BC) in Upper Egypt.[21] In the Naqada culture (3500–3200 BC), copper replaced stone as a material for axes, blades, jewelry, rods, and spatulas. Large amounts of copper ore were found at Maadi, in Lower Egypt.[22] It was probably transported from the mines in the Sinai Desert that bear inscriptions naming Djoser, Sneferu, and Khufu. In later periods, the Egyptians made copper statues and even copper drain pipes.[23] We can assume that Fourth Dynasty builders had the know-how and resources to make all the copper tools required for construction projects.

Tools

Laborers used tools resembling mattocks or hoes for digging and woven reed baskets for hauling dirt and debris. Stonemasons used copper chisels and saws, drills, wooden mallets, wedges, wooden rollers, stone hammers, and smooth, round balls of dolerite, a hard stone used to cut blocks of granite.

A wide variety of woodworking tools existed, including mallets, hammers, drills, copper saws, chisels, and scrapers or planes.

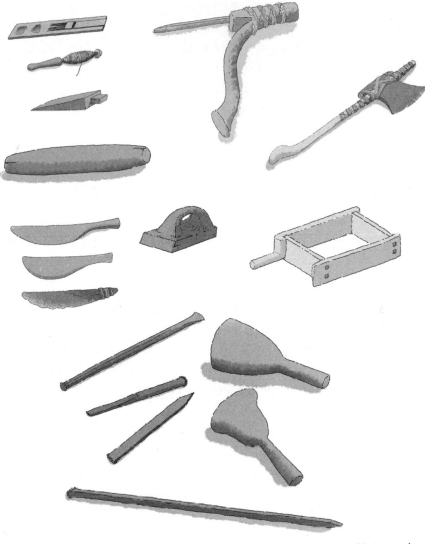

Various tools.

Many copper tools found at pyramid sites are displayed in the Egyptian Museum in Cairo and other institutions. These include assorted copper chisels—one 48.3 centimeters long—as well as a copper plasterer's trowel. It is apparent from saw marks on Khufu's sarcophagus that the ancient Egyptians used saws capable of cutting granite that were at least 2.44 meters long. These saws may have been used with an abrasive powder. I've seen saw cuts in the basalt floor-paving stones at Khufu's mortuary temple. Lehner reports that traces of copper and an abrasive can be found in these cuts.[24] Holes drilled in hard stone with diameters ranging from a few millimeters up to 50 millimeters have been reported. As seen in tomb paintings, these drills were weighted to increase the bearing pressure and turned by hand.[25] Drills were also used with an abrasive powder for cutting.

TRANSPORT OF MATERIALS

On land, the ubiquitous donkey was used to transport goods. However, the main artery for transportation was the Nile River. Skilled boat-builders and sailors, the Egyptians developed vessels ranging from small reed boats to oceangoing vessels. They sailed the Mediterranean coast-line as far north as what are now Palestine, Lebanon, and Syria and crossed the Red Sea to Saudi Arabia.

For transporting stone, they used sailing vessels and barges, some with capacities of at least 100 metric tons and perhaps as much as 1,000 metric tons. For larger loads, it is likely that rafts made of logs were assembled. Clarke presents an overview of Egyptian boats.[26]

ERECTION TECHNIQUES

No wheels are known to have existed during the Fourth Dynasty, and the builders of Khufu's pyramid did not have wheels or pulleys for moving or lifting loads. Instead, blocks of stone were levered onto wooden sledges and then dragged to the construction site by teams of laborers. Wooden levers were used to move loads over short distances, but for longer hauls it was necessary to construct ramps to bring blocks of stone up to the higher levels of the structure. Alternatively, stones could be tumbled or rolled for short distances by pulling them with a rope harness while another laborer used a lever to raise the rear of the stone. For stone placement, a block could be set on a round dolerite stone (much like a ball bearing) and maneuvered into place. For placing very large backing stones, raised bosses were left on the base for levering, and notches for levers were cut in the foundation platform or adjacent stones. These were later filled in or covered.

To place heavy roof beams, the room was filled with sand, and workers placed the beams (typically granite) while supporting them on the sand base. Then the sand was excavated. A similar approach was used to lower massive stones or a sarcophagus into a subterranean chamber, placing it on the sand and then removing the sand.

A PLAN UNFOLDS

The means and methods of constructing Khufu's pyramid clearly were available and at the disposal of Hemiunu. This capability came about as a result of a century-long evolution in construction technology, from the large mastabas to the first step pyramid and finally to the true pyramid.

As the de facto program manager, Hemiunu's next step in overseeing the building of the pyramid would have been to recruit a small cadre of expert builders to lead the work. In all likelihood, he was confident he could secure their commitment. After all, he was offering them a role in the greatest construction project the world had ever known. No doubt he knew his men. None of them would be able to resist such an opportunity. By modern standards their reward may seem simple or insignificant. But when the structure was complete and towered above the surrounding landscape, they knew in their hearts that it was the result of their efforts. In unspoken confirmation of their work, they found ways and places to conceal their names on stones deep inside the pyramid. Their graffiti, lost to the ages until recently, made their names immortal.

CHAPTER 3

A TOMB FOR A KING

The Great Pyramid at Giza is the best known and largest of nearly 100 pyramids constructed along the west bank of the Nile (over a distance of roughly 100 kilometers) during the Third to Sixth Dynasties in the Old Kingdom of Egypt. Also known as the Great Pyramid of Khufu (or Cheops, as the Greeks called him), it was built during the Fourth Dynasty, about 2,550 years before the birth of Christ. Khufu reigned from 2551 to 2528 BC.[1] The square base of the pyramid covers an area of 53,000 square meters. Originally the pyramid was 146.6 meters high, but the top 8 meters have disappeared. The volume of material contained in a pyramid with these dimensions is 2.59 million cubic meters. The Giza plateau is the site of not only Khufu's Pyramid, but also those of his son Khafre and his grandson Menkaure—and of several smaller pyramids known as the Queens' Pyramids.

No records have been found that explain the move from Dahshur, where Khufu's father built the Bent and Red Pyramids, to Giza. It may have been that the supply of good, readily available stone had been exhausted at Dahshur, or the builders wanted a rock site better able to support the weight of the pyramid, or perhaps Khufu wanted to escape being in the shadow of his father's great works. Whatever the reason, we can

Khufu's pyramid seen from the east face.

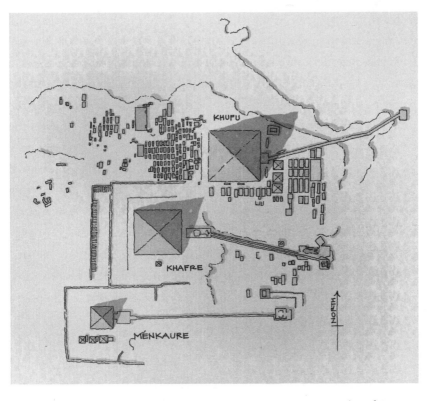

Plan of Giza site.

assume that a group of experienced builders and stonemasons were dispatched to locate a new site along the west bank of the Nile River. They sought a site with ample stone for building and with underlying bedrock capable of supporting a massive structure. The Giza plateau was ideal.

Once the site was identified, it was visited by priests and consecrated for its holy purpose. One can imagine the reaction of the first visitors to the new site. Few plants grew there, and there was no water except for the river in the distance. There were no roads, not a single house within 100 khet (5.2 kilometers) in any direction. Everything would have to be planned for and built to construct the pharaoh's tomb. A mammoth undertaking lay ahead.

With the site selected, Hemiunu summoned his most experienced planners, designers, builders, and artists to a meeting. There he outlined the broad parameters of the project for them, based on information collected from earlier meetings with the foremen, supervisors, artists, and designers who worked on the Red Pyramid for Khufu's father. He may have had a model carved from limestone or fine white alabaster, or a sketch, or he may have just spelled out the design goals and left it to the planners, architects, and builders to determine how best to meet the project goals: the height was to be 280 cubits; the width and length of the base, 440 cubits. After a suitable pause to let the meeting participants grasp the concept—that this was to be bigger than anything previously built in Egypt, and by a large margin—Hemiunu directed the meeting toward the practical considerations of getting the job done. A new quarry would have to be surveyed and laid out. A harbor would be constructed and access roads built from the quarry and the harbor to the site. Housing for workers, workshops, and storerooms had to be provided, and plans had to be made for all the other facilities needed to support a workforce of thousands of people.

At Giza, the builders were fortunate in having an ample supply of high-quality limestone. The stone lay 30 meters thick—enough to build three pyramids, temples, and other structures.

To gain a better appreciation of the challenge that Hemiunu undertook, consider the sheer physical size of Khufu's pyramid in comparison to a large modern structure, Hoover Dam. Completed in 1936 after five years of construction, it was for many years the largest masonry dam ever constructed. The concrete base of Hoover Dam is 201 meters thick, not quite as thick or wide as the base of the pyramid. The dam is 379 meters long, about 60 percent longer than the pyramid, and 221 meters high, or about 50 percent higher. The dam contains 2.48 million cubic meters of concrete, less than the amount of limestone in the pyramid. It took five years for an average workforce of 3,500 people working full-time to build it (the peak workforce was 5,218). The great height of the dam and the narrow canyon in which it was placed presented special challenges to the construction workforces.

The advantages at the disposal of the Hoover Dam workforce— mechanized equipment, trucks, cranes, and other power tools—were offset by the added complexity and difficulty of the project. For example,

Hoover Dam

the first step in building the dam was to divert the Colorado River through tunnels blasted in rock. Then cofferdams were constructed upstream and downstream, and the site was pumped dry so a foundation could be built. This required the excavation of more than 380,000 cubic meters of silt and mud and 760,000 cubic meters of rock. More than 4.4 million cubic meters of material were excavated. This work went on twenty-four hours per day. The complete project cost $165 million. The highest-paid laborers, who set explosives on the canyon walls, received $5.60 per day. Ninety-six workers lost their lives constructing the dam.

In contrast to Hoover Dam, which was constructed in a narrow canyon, Khufu's pyramid was constructed on a rock bluff. After clearing the Giza site, the Egyptians leveled most of the base area, with the exception of an outcropping of rock that rises above the plain.

Imagine the immensity of the task faced by Hemiunu, who had to build a structure comparable to Hoover Dam, but with no machinery, no pulleys, no wheels, and no iron tools. Large blocks of limestone and granite—some weighing as much as 50 metric tons—had to be cut at the quarry established at the site, or in some instances at distant quarries. The local limestone was then transported from the quarry to the plateau. After the masons cut and trimmed the stone, it was levered onto a wooden sledge and dragged up a sloping earthen ramp to the work site. Examples of such sledges can be seen today in the Egyptian Museum in Cairo. An illustration of one, used to transport a statue of a princess, was observed in a tomb near Khafre's pyramid.[2] The remains of earthen ramps are evident at Giza and a number of other pyramid sites I inspected. The stones from distant quarries were also moved to the shore of the

Nile and transported across or down the river by boat to the site. The rock used to construct the pyramid core was carved from a quarry at the site (between the Sphinx and the pyramid), but the white limestone used as the exterior casing of the pyramid came from a site called Turah, across the Nile River. Even today, limestone is produced in this area.

The builders of Khufu's pyramid waited to transport materials until the flooding of the Nile brought water to within about a quarter mile of the site.[3] Blocks of limestone weighing anywhere from 1 to 5 metric tons made up the bulk of the pyramid. A little more than 2 million such blocks were cut from the quarry at Giza, near the pyramid site. Approximately 100,000 blocks of white limestone casing stones from Turah were added to finish the construction.

Egyptian workmen perfected the technique of using hand-driven copper chisels or drills to cut slots or holes in the stone faces. After a slot (just wide enough for a man to stand in sideways) had been chiseled out along each side of a block, holes were cut at the bottom edge, wedges were inserted into them, and the block of stone was broken loose by pounding on the wedges with mallets. These rough blocks were dressed down to finished dimensions. On the exterior faces of the pyramid, the final dimensions of these stones were extremely accurate, so the joints could be made in fractions of a millimeter—in some cases less than 0.5 millimeters. Less care was taken with the interior blocks: gaps in the masonry were filled with chips, rubble, small stones, or mortar.

The pyramid was oriented with its major sides north–south and east–west. This in itself was a remarkable undertaking, given the accuracy with which it was done, because the ancient Egyptians had no compass and had to use astronomical observations.

There is no conclusive evidence of exactly how the Egyptians lifted the stones and placed them on the pyramid. A number of theories have been advanced through the ages. Herodotus, writing 2,500 years ago (2,000 years after the pyramids were built), indicated that a system of levers was used.[4] It has also been suggested that long wooden poles (in the form of a machine resembling a giant seesaw) were used to elevate the blocks from one level of the pyramid terrace to the next—either by using multiple levers or by moving the levers themselves to each higher elevation as it became necessary. Modern attempts to duplicate this method ended in failure.

Using this method to place such a large number of stones within the lifetime of the pharaoh would have required a large quantity of strong timbers. Since timber was not readily available in Egypt, this approach does not seem practical, and to my knowledge virtually no examples of timber used in pyramid construction have been found—with the exception of the three wooden beams that I observed protruding from the east face of the Step Pyramid at Saqqara. If timber had actually been employed on the scale necessary for the lever method, more wood would have been found in the structures. Excavations by Mark Lehner and Zahi Hawass suggest that the limited wood available was used for cooking, in smelters used to make copper for tools, and possibly to make a poor-quality lime or gypsum cement used in mortar. Some timber was certainly used to build sledges, for rollers, for levers to pry stones out of the quarry, for temporary shoring, for maneuvering stones into position, and for the scaffolding used by artists and sculptors, which may have been used in finishing the exterior of the pyramid. (See Plate 13.)

In recent times a number of exotic—some would say preposterous—construction schemes have been proposed. I could not find any convincing technical merit in them.[5]

There is *considerable* evidence, however, supporting the use of an inclined ramp or series of ramps. We know that sloping ramps were used in the construction of other pyramids (see Chapter 6). But it is unlikely the Egyptians built a single ramp, because a ramp that reached to the top of the pyramid would have been nearly a kilometer long and contained more material than the pyramid itself.

After the site was leveled, an initial course of blocks was placed to outline the base of the pyramid. This placement was executed with extreme care because it formed a reference point for the other dimensions as construction proceeded. At this point construction of the Descending Corridor began, and openings were left for the Ascending Corridor, which would provide access to the chambers within the pyramid itself.

INTERIOR FEATURES

The drawing below shows details of the inner structure of Khufu's pyramid. Each of the principal internal structures posed considerable construction challenges. Entering the Descending Corridor, one can see the joints in the masonry used to frame the passageway, which is composed of blocks roughly 1.5 meters long and 1.2 meters high that form the walls and ceiling. The upper portion was clearly constructed as the pyramid was being built. The aboveground portion of the tunnel rises through the body of the pyramid and emerges from the north face at a vertical elevation of around 17 meters, or at what I call Course 19. About one-third of the way down, the corridor branches upward, leading to the Queen's Chamber and the King's Chamber, high up in the center of the pyramid, while the passageway continues downward. (See Plate 15.)

I believe that, prior to laying the foundation course, Hemiunu's surveyors marked the spot where the Descending Corridor would be cut. It was to angle down below ground at a slope of 26.5° to provide access to a subterranean room known as the Lower Chamber. Below the sur-

A sectional view of Khufu's pyramid, showing chambers and corridors.

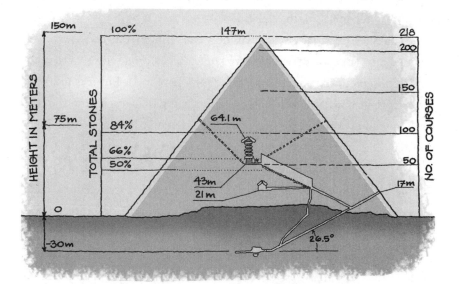

The Descending Corridor of Khufu's pyramid.

face, the corridor becomes horizontal, terminating after a short distance in the Lower Chamber. The corridor, a little more than 1 meter wide, is only 1.26 meters high and decidedly uncomfortable to descend. It is necessary to walk stooped over, at constant risk of banging one's back or head on the rough stone above. While the stonecutters labored underground, construction would have proceeded on the surface, allowing the miners' debris to be removed from below.

The Lower Chamber was never completed. It was hewed out of bedrock in a rectangular shape, but one area is unfinished. Channels in one corner—barely the width of a person and cut into the limestone— were made to prepare for hewing out additional blocks of stone to enlarge the chamber. A stonemason stood in this narrow channel, one foot in front of the other, and hacked at the limestone with a mattock or hammer and copper chisel. After cutting a channel around 15 centimeters wide and 1 meter deep on three sides of a block, the masons would insert levers and break out a piece of stone that would then have to be broken down further to maneuver it laboriously up the Descending Corridor. The most difficult part must have been the initial cut at the ceiling line. There, the mason had to lie on his stomach in a narrow groove, working his tools horizontally, chips and limestone dust dropping in front of his face. Looking at the tool marks and rough cuts that remain, one can only marvel at the fortitude of the stonecutters who performed this hard, dusty work in an area with little light and ventilation.

Features in the Lower Chamber suggest it was planned to be more complex. There is an unfinished recess in the west wall, as well as a pit in the floor. From the south wall, a small, cramped tunnel extends some 5 or 6 meters horizontally to a dead end. Here again, we must admire the tenacity of the stonemasons who crawled on their stomachs, chiseling this narrow opening through solid bedrock. A shoulder-wide tunnel 30 meters below the surface of the earth was no place for the claustrophobic!

The purpose of this unfinished tunnel and chamber is not clear. Was it originally intended as the burial chamber? We will never know for certain. What is known is that the stonecutters laboriously removed 430 cubic meters of rock to create it. This rubble was probably used as fill in the lower layers of the pyramid to minimize the backbreaking labor required to remove it.

Meanwhile, above ground, masons leveled the foundation platform and began laying the first few courses of stone. At some point the Lower Chamber was abandoned and the decision reached to build a chamber for the pharaoh's final resting place above ground in the upper portion of the pyramid. From the Descending Corridor, the Ascending Corridor was constructed, rising to the center of the lower portion of the pyramid. Edwards reports that a hole was cut in the roof of the Descending Corridor about 28 meters from the outside entrance, accessing a corridor that rises at an angle of 26.2°.[6] The lower portion of the Ascending Corridor is framed with special stones, called "girdle stones," placed every 5.2 meters (see Chapter 7). The 39-meter-long Ascending Corridor rises to a height of around 21 meters (corresponding to Course 23) and then becomes horizontal and reaches a point midway between the north and south sides of the pyramid. Here another chamber was

The Horizontal Corridor,
leading to the Queen's Chamber.

constructed, known today as the Queen's Chamber. Currently, the Ascending Corridor is reached through a passageway originally opened by the grave robbers who first gained access to the pyramid. Continuing down this passageway, one enters the Ascending Corridor at a point past the blocking plugs, which will be described later.

The Queen's Chamber is a misnomer, because this chamber was most likely intended for Khufu. (It was mistakenly given this name centuries ago by Arab explorers.) The room itself measures 5.75 by 5.23 meters and has a vaulted ceiling that rises to a height of 6.22 meters. The roof of the Queen's Chamber has six large beams on each side, twelve in all, which form a V-shaped ceiling—technically, a pointed saddle roof. The walls and ceiling have tight joints and a fine finish. On the east wall, a recessed alcove with a corbelled ceiling might have been intended for a statue of Khufu. This alcove has a depth of about 1 meter and is framed in blocks about 1 meter high. Two such blocks, one on top of the other, form the vertical sides of the alcove, and the next four blocks are each offset about 13 centimeters toward the center to form the corbelled ceiling. In this manner the width of the alcove, 1.6 meters at the lower portion, decreases to 0.5 meters at the top. There are also two square shafts in the walls of the Queen's Chamber, one facing the entrance tunnel and one on the opposite side—that is, on the north and south sides. They are about 20 centimeters square and exit the chamber at an angle of about 30+°. While there has been much speculation about these small shafts, their purpose is not known. The floor of the Queen's Chamber is in a rough condition, indicating that this chamber was never finished. A logical approach to building the Queen's Chamber would have been to erect the pyramid to the level of its floor (approximately Course 23), thus providing a level platform for a working surface. The walls would have been erected, the chamber backfilled with sand, the ceiling beams put in place, and then the sand removed. (See Plate 9.)

Above the Queen's Chamber is a large, unique gallery that is basically an extension of the Ascending Corridor but can be entered without stooping over. The Grand Gallery is a truly spectacular example of stonemasonry. In cross section it resembles the Queen's Chamber alcove, but on a much larger scale. The vertical walls rise about 2 meters, and the width at the base is likewise about 2 meters. Interestingly, this is close to 4 by 4 cubits, probably by the designer's intent. At the foot of each ver-

The Grand Gallery.

PLATE I. The only known likeness of Khufu is this small ivory statue.

PLATE 2. A mastaba at Meidum.

PLATE 3. The pyramids at Giza viewed from the southwest.

PLATE 4. Djoser's Step Pyramid at Saqqara.

PLATE 5. The ruins of the pyramid at Meidum.

PLATE 6 The Bent Pyramid at Dahshur.

PLATE 7. The Red (North) Pyramid at Dahshur.

PLATE 8. Chamber with corbelled ceiling in Red Pyramid.

PLATE 9. The Queen's Chamber, showing the alcove.

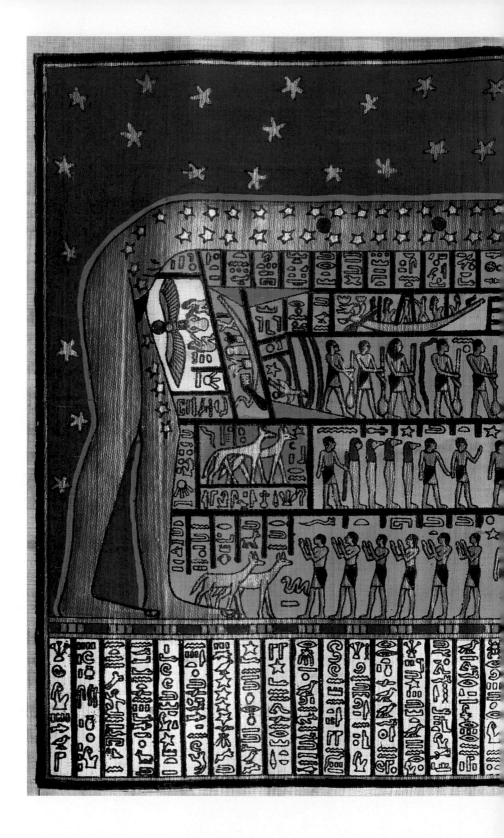

PLATE 10. The Goddess Nut.

PLATE 11. Khufu's pyramid seen from the northwest corner, showing the courtyard and a few remaining casing stones (lower course, right side).

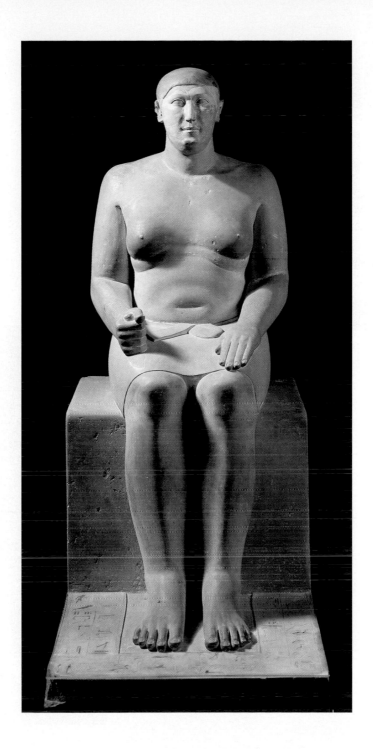

PLATE 12. Hemiunu's statue.

PLATE 13. A tomb painting of men working on a scaffold.

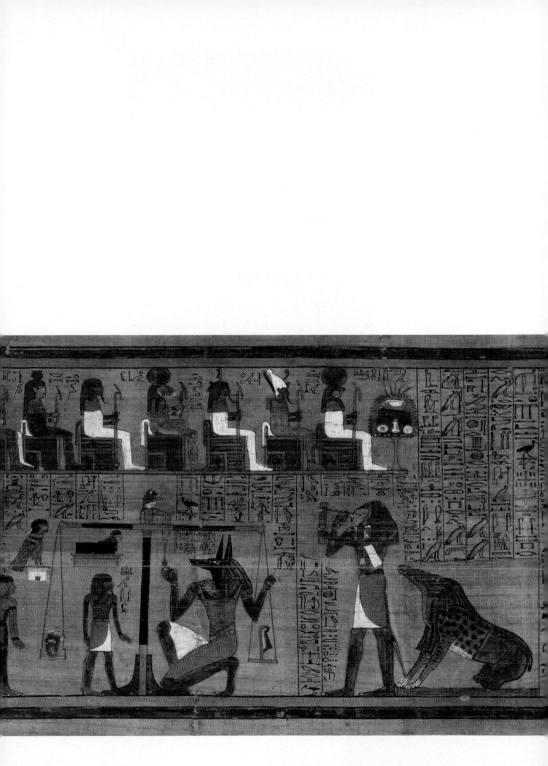

PLATE 14. Weighing the heart to enter paradise.

PLATE 15. The original entrance to Khufu's pyramid.
Note the supports over the opening.

PLATE 16. Ramps and causeways at Giza and other sites.
TOP: Khafre's pyramid causeway, right background; old quarry, left center; Menkaure's causeway, left foreground. BELOW LEFT: Ramp to mastaba, west side of Khufu's pyramid, Giza. BELOW CENTER: Ramp retaining walls, south of Khufu's Queens' Pyramids, Giza. BELOW RIGHT: Causeway at the pyramid of Unas, Saqqara.

PLATE 17. The Wall of the Crow.

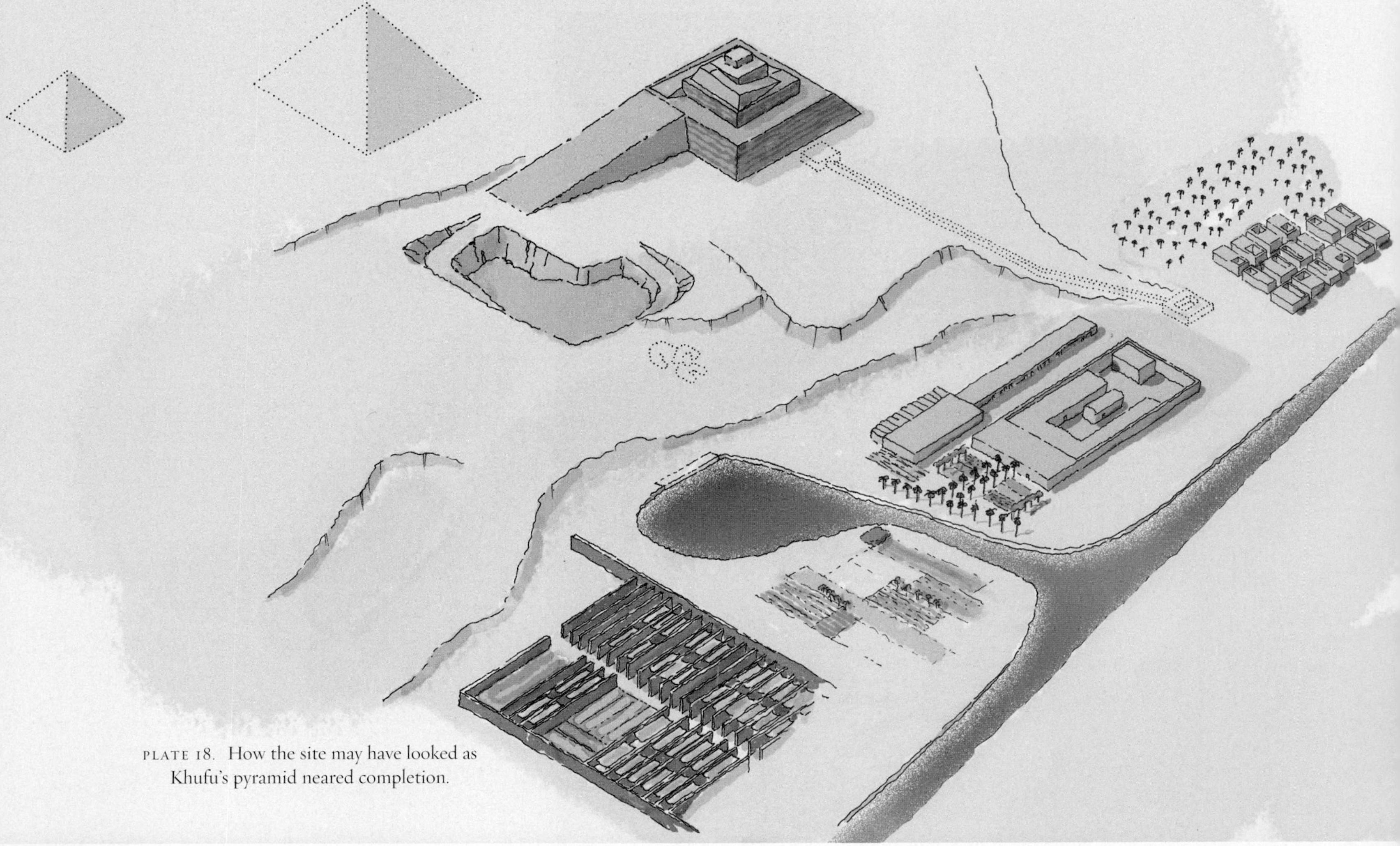

PLATE 18 How the site may have looked as Khufu's pyramid neared completion.

tical wall, a stone curb extends 50 centimeters into the passageway. In the depressed center space between the two curbs, a wooden walkway has been fitted with steps and handrails for the convenience of visitors. Above the vertical walls are seven steps to the corbelled ceiling. The lower steps (those I could reach to measure) are 84 centimeters high. The overall height of the gallery is 8.6 meters, and its total length is 46.1 meters.[7] The corbelled ceiling does not meet at the top of the Grand Gallery. Instead, it is bridged by roofing slabs with a clear span of two cubits, the same distance between the curbs on the floor.[8]

The Grand Gallery is one of the most remarkable examples of the ancient Egyptians' application of a corbelled ceiling. It has several other interesting features. In addition to the flat curbs or ramps running along each side wall for the entire length, there are twenty-seven niches and slots cut into the walls. Their exact purpose is not known, but they may have supported a scaffold or wooden support structure, or enabled the Ascending Corridor to be sealed after the pharaoh's body had been laid to rest. Various theories have suggested how the three granite plugs that seal the Ascending Corridor were pulled up into the Grand Gallery for storage. A wooden structure could have been placed over them to allow the funeral procession to move up through the Grand Gallery. After the interment ceremony, workers could have released them from behind, sealing the entrance. They would have then exited the pyramid through the vertical shaft that connects the end of the Ascending Corridor (just before the entrance to the Grand Gallery) with the Descending Corridor. Then, as the final step, the entrance to the Descending Corridor would have been sealed and the final casing stones put in place to cover and conceal the entrance.

At the end of the Grand Gallery, a short horizontal passage leads into the King's Chamber through a small opening. A portion of this passage is constructed of red granite. Here we find an antechamber—or more properly, a portcullis—that appears to have been designed in such a way that the chamber could be sealed by blocks of granite slid into place after the burial. There are four sets of slots in the side walls of this room, and clearance overhead, so heavy slabs of granite could have been slid in from above (before the roof was placed) and propped up on supports in some manner. After the king's remains were placed in the sarcophagus, the supports would have been removed and the slabs lowered by ropes, effectively blocking access to the King's Chamber.

The King's Chamber was intended as the final resting spot for Pharaoh Khufu. Entrance is gained through a small opening, roughly 1 meter wide by 1 meter high. A granite sarcophagus is situated near the west wall. One corner is damaged, possibly by grave robbers who entered the pyramid and pried the lid off. The sarcophagus is wider than the Ascending Corridor, indicating that it was placed in the chamber and the walls raised around it. The chamber is constructed of red granite, its walls rough-polished. The internal dimensions are 5.23 meters wide by 10.5 meters long by 5.8 meters high.[9] The walls are five courses high, each course a little more than 1 meter high. The flat ceiling is constructed of nine massive granite beams, each weighing 40 metric tons or more.

The King's Chamber and sarcophagus.

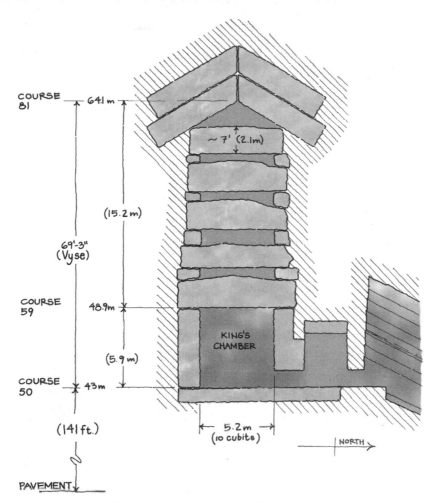

COURSE
81

64.1 m

~ 7' (2.1m)

(15.2 m)

69'-3"
(Vyse)

COURSE
59

48.9m

KING'S
CHAMBER

(5.9 m)

COURSE
50

43m

(141 ft.)

5.2 m
(10 cubits)

NORTH

PAVEMENT

A sectional view of the Relieving Chambers, Khufu's pyramid. After Vyse (1840), 158.

In the King's Chamber, as in the Queen's Chamber, small shafts rise to the north and south faces of the pyramid. Those leading from the Queen's Chamber do not reach the exterior of the pyramid; those leading from the King's Chamber do. Their exact purpose is not known, but it is surmised that they are ceremonial exits allowing the pharaoh's spirit access to the heavens. Recent efforts have been made to explore these shafts with a miniature, remotely controlled vehicle bearing a television camera. A small door was observed in one.

Above the roof is a structural feature that has not been observed in any of the earlier pyramids: a series of five compartments called relieving chambers, one above the other, that relieve the load on the roof of the

King's Chamber. The uppermost is distinguished by a pointed roof. The rough limestone walls still bear the red ocher quarry marks that guided the masons who cut them. There is also graffiti referring to Khufu left by gangs of workmen.

As construction reached the level of the antechamber and King's Chamber, they were undoubtedly built in place, slightly ahead of the rising terraces of the pyramid. Thus it was possible for the workmen to install the finely dressed walls, lintels, and ceiling blocks of these interior spaces from a level surface and then build the rest of the pyramid up around them as construction proceeded. This approach provided a flat working surface for maneuvering and placing the large granite roof beams and other heavy stones used in the King's Chamber and the Relieving Chambers above it.

ANOMALIES AT HIGHER ELEVATIONS

One of the striking features about the construction of Khufu's pyramid is that the height of the courses of masonry in the pyramid is not uniform. As the builders proceeded upward, the height of the exterior casing stones decreased. To illustrate this trend, I measured the height of the first nine courses or levels at a point near the midpoint of the east face.

Height of Courses

COURSE NO.	HEIGHT (CM)
0 (platform)	53
1 (first layer of backing stones)	149
2	122
3	110
4	107
5	100
6	97
7	86
8	85
9	90

Because of weathering, these measurements easily could vary by $\pm$ 1 centimeter, depending on where the measurement was made, but the trend is unmistakable. Course 0 is the 1-cubit-thick base platform, which was carefully measured and prepared to provide a level base for construction. Course 1 consists of very large stones that delineated and fixed the base of the structure. A stone in Course 1 had these dimensions: height, 149 centimeters; width, 212 centimeters; length, 230 centimeters—a volume of 7.27 cubic meters and a weight of 16,000 kilograms. Each course becomes progressively smaller. For example, the stone in Course 9 had these dimensions: height, 90 centimeters; width, 80 centimeters; length, 100 centimeters—a volume of 0.72 cubic meters and a weight of 1,584 kilograms.[10] This trend continued up to Course 19, where once again there is a layer of blocks about 1 meter high, and then again to Course 35, where there is a layer 1.26 meters thick.

There are somewhere between 210 and 220 courses of masonry from the platform level to the top of the pyramid. W. M. Flinders Petrie, in his classic work, provides measurements of the height of each course at several locations, including the northeast and southeast corners.[11] His measurements stopped at Course 203 (elevation 138.6 meters) on the northeast corner. Since the height of the pyramid can be established as 146.6 meters, the top section of about 8 meters is missing.

Petrie's published data contain a few discrepancies, however. In several cases it appears he misread his handwritten measurements or incorrectly added the incremental heights when calculating the height at certain courses. Also, his thickness measurements vary with location, suggesting that each course is not precisely leveled at every point. However, there are repeated instances in which the variance in elevation diagonally across the flat surface is 0 to 2.5 millimeters, which suggests that the builders took special care to level the working surface every ten courses on the average.[12] It is also significant that the carefully leveled courses correspond to the levels where important internal structures were constructed—that is, Course 19, the main entrance; Course 23, the floor of the of the Queen's Chamber; Course 50, the floor of the King's Chamber; and Course 80, the top of the Relieving Chambers.

I spot-checked Petrie's data against my own measurements of the platform and the first nine courses, as listed above. My measurements, taken near the center of the east face, agree fairly well with his data for the northeast corner, considering that more than 100 years of weathering and tourist abuse have passed since Petrie made his measurements.

To estimate the number of courses, I averaged Petrie's data for the northeast and southwest corners and assumed the last courses were the same height as Course 202 and that the pyramidion, or capstone, was 0.9 meters high, which gave 218 courses to the top. Using the same approach with the *maximum* values Petrie measured, I determined that there were 210 courses to the top. In this second model, about 165,000 fewer stones are required. The difference comes about from using larger blocks. For the purposes of this book, I used Petrie's averaged data and a hypothetical figure of 218 courses. (Details are provided in Chapter 6.)

The graph opposite shows the trend in block sizes as a function of course number, based on Petrie's data. The block height trends downward for a number of courses, jumps back to a higher value, and then repeats the same pattern. The maximum block heights occur at Courses 1, 7, 19, 35–36, 44–45, 67, 74–75, 90–91, 98–99, 118–19, 138–39, 144, 164, 180–81, and 196–97. Evidently, there is a ring of larger backing stones (1.0 to 1.3 meters high at the lower levels) every fifteen to twenty courses. In some courses blocks of double thickness overlap two courses and provide added structural stability. These larger blocks serve to strengthen the structure and provide a means of correcting any misalignment in the masonry.

At the higher levels (above Course 80), the average height of the blocks decreases to about 60 centimeters, and above Course 165, to about 50 centimeters. For the last eighty courses or so, the nominal height of most courses is one cubit, ± 10 percent. Assuming that the same proportionate height:width:length ratios prevail, the blocks at the uppermost levels weigh between 0.75 and 1.0 metric tons. Less labor is required to transport and place the stones at the higher levels of the pyramid.

I used Petrie's data to construct a mathematical model to calculate the total number of stones and vertical height to each course, and to

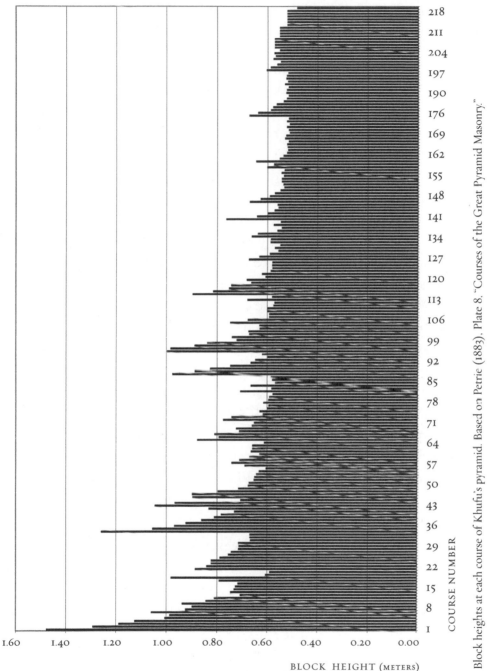

COURSE NUMBER

218
211
204
197
190
176
169
162
155
148
141
134
127
120
113
106
99
92
85
78
71
64
57
50
43
36
29
22
15
8
I

1.60 1.40 1.20 1.00 0.80 0.60 0.40 0.20 0.00

BLOCK HEIGHT (METERS)

Block heights at each course of Khufu's pyramid. Based on Petrie (1883). Plate 8. "Courses of the Great Pyramid Masonry."

develop the pyramid cross-section drawing on page 92. This model is useful for the insight it provides into the logistics of construction. For example, by Course 55, one half of the stone blocks were in place.[13] Course 23 provided a flat surface upon which the Queen's Chamber could be constructed and work on the Grand Gallery could commence. Course 50 provided a base for constructing the King's Chamber. Once the massive beams that make up the Relieving Chambers, above the King's Chamber, were put in place (Course 80), there was no more need for a large ramp. The remaining stones typically weigh 1,000 kilograms or less, and a smaller team of laborers could transport and install them.

FINISHING THE PYRAMID

The outer courses of the pyramid were constructed of limestone backing stones, which were cut and fitted more accurately than the rough core stones. Once the last course of backing stones was in place and the pyramidion had been installed, the ramp and scaffolding were removed to expose several courses of the backing stones. Then the finishing layer of white limestone casing stones from Turah was placed over the backing stones to give the pyramid its smooth exterior surface. It is possible that the topmost casing stones were precut and finished on the ground and then placed at the apex of the pyramid. The others were installed by working downward, as the ramp and scaffolding were removed. The edges of these stones were trimmed to the precise angle of the pyramid (51.9°), but the stones themselves would have been left with excess material to be trimmed in place, ensuring very fine joints and a seamless surface. Examples of casing stones prepared in this manner can be seen on Menkaure's pyramid, and a few remain on the north side of Khufu's pyramid.

Unfortunately, we have to resort to our imaginations to visualize what Khufu's pyramid looked like when it was fully sheathed in dazzling white Turah limestone. The exterior finish has been removed, except for a few pieces near the base of the pyramid. Modern historians speculate that the white limestone was stolen over the millennia to build bridges, houses, and terraces in Cairo and the surrounding area. Today, all the visitor can see is the weathered stair-step appearance of the backing stones, although segments of the white limestone terrace that surrounded the pyramid have survived.

OTHER FEATURES OF KHUFU'S PYRAMID COMPLEX

Next to Khufu's pyramid are three smaller satellite pyramids, today referred to as the Queens' Pyramids, which may have once held the remains of various queens. The current thinking is that the northernmost one (GIa) was intended for Queen Hetepheres (Khufu's mother), and the others were for his wives—the center one (GIb) for Queen Meretites, and the southernmost one (GIc) for Queen Henutsen.[14] Royal women did not share the King's Chamber, but they were accorded their own special burial chambers, albeit in small replicas of the king's. Perhaps this reflects the pharaoh's understanding of conjugal affairs in the afterlife, best expressed in a line from a seventeenth-century poem by Andrew Marvell: "The grave's a fine and private place, / but none, I think, do there embrace."[15]

The Queens' Pyramids were not constructed with the same precision or care as Khufu's pyramid. They are basically step pyramids of three steps covered with casing stones to effect a smooth pyramidal finish. The sides of the base are approximately 46 meters long, which makes them one-fifth-scale replicas of Khufu's pyramid. (With an angle of 51.9°, they have a height of 29.33 meters, or one-fifth the height of Khufu's pyramid.) The burial chambers are all beneath the pyramid in bedrock.

Although Queen Hetepheres' remains were never found, in 1925 her burial assembly (furniture, disassembled canopy bed, carrying chair, pottery, disintegrated linen cloth, and other items) was found in a deep shaft tomb near the northeast corner of

Khufu's Queens' Pyramids.

Queens' Pyramid G1a. The archaeological evidence indicates that this was a reburial. Also near this area is a descending corridor and an ascending corridor cut in the bedrock. Some researchers believe these were "trial" passages for the corridors in Khufu's pyramid. However, Lehner establishes a strong case that this was the original location of Queens' Pyramid G1a and that subsequently the planners decided to move it west to its current location.[16]

Adjacent to Khufu's pyramid are several carefully constructed rock pits that contain—or once contained—disassembled boats to convey the pharaoh on his journey to the afterlife. Two of these were found to have boats in them; one was opened and the boat carefully reassembled and displayed in a museum on the site. The size of this vessel (43 meters long, with a beam of almost 6 meters) and its displacement (45 metric tons) provide ample evidence of the Egyptians' ability to construct vessels or barges capable of transporting the raw materials needed to construct the pyramid.

Although the pyramid is the most visible and dramatic remnant of Khufu's burial site, the entire complex contained several other important features. Only traces of these remain, but it is possible to get a good idea of what they looked like by considering the more complete remains found at Khafre's pyramid. The entire pyramid was enclosed by a wall constructed of fine Turah limestone, all of which is now gone. The area around the pyramid was paved with limestone, vestiges of which remain.

Facing the east wall of the pyramid was a mortuary temple. This limestone building is believed to have been 52 meters long and 40 meters wide. Traces of the black basalt floor stones remain. Granite pillars supported a roof structure.[17] The mortuary temple was connected by an elaborate causeway to another temple below the plateau, roughly at the elevation of the Nile floodplain. Only traces of the upper portion of the causeway can be seen; the remains of the lower portion, and indeed the site of Khufu's valley temple, are now hidden from sight by the encroachment of the modern city.

What was Khufu's valley temple like? We can only imagine, based on the remains of Khafre's valley temple. It must have been impressive. I visualize Hemiunu strolling through the valley temple, up the causeway, and finally entering the mortuary temple, perhaps to nervous protests from attendant priests for intruding in their exclusive domain. Finally, he emerges from the temple to the white limestone courtyard at the base of the pyramid, and looking upward, he says silently to himself, "I built this."

CHAPTER 4

WE CONSECRATE
THIS HOLY GROUND

The Egyptian builders faced an enormous challenge in preparing the desert site for the workforce that performed the construction. The site was selected because of its prominence (it occupies a plateau overlooking the Nile River), its abundant raw material, and because it provided ideal bearing support for a massive structure—the necessity of which was one of the most important lessons learned from building the earlier pyramids at Saqqara and Dahshur. It was close to the ancient city of Memphis, the capital at that time. Khufu no doubt had concerns—not only about site conditions at Dahshur, but also about the fact that the best and most accessible limestone had already been consumed there. It is also quite conceivable that Khufu did not want to build his pyramid in the shadow of the two pyramids erected at Dahshur by his father.

It is impossible to imagine that a public works project on the scale of Khufu's pyramid would not leave traces of the construction process on the landscape. During my own explorations of the site I observed clear evidence of the ancient quarries—some containing blocks marked and half cut out, waiting for the stonemasons to return and remove them—and there is unmistakable evidence of ramps and ramp debris that was dumped into these quarries. Lehner systematically explored facilities that would have supported a large local workforce, and Hawass explored the graves of these workers. The remaining missing elements are the harbor that permitted ship transport from Aswan and Turah, and the supply road that connected the harbor to the construction site. Lehner presents compelling evidence of these two remaining features.

Lehner and his colleagues conducted an extensive mapping of the Giza Plateau, and on the basis of this data and historical maps and photographs, they prepared topographical maps of the site before, during, and after the construction of the pyramids.[1] The Giza Plateau is an outcrop of a middle Eocene limestone formation called the Mokkatam. It extends 2.2 kilometers from east to west and half that distance from north to south, dipping 5–10° toward the southeast. The elevation at Khufu's pyramid is roughly 60 meters above sea level, and the low point of the best limestone is between 20 and 30 meters above sea level, so there are about 40 meters of thickness to mine for building materials.

The Sphinx stands on the eastern edge of one of the quarry areas and is believed to have been carved from a rocky outcropping that remained following quarry operations. The top of the Sphinx's head is postulated to be at the level of the original ground surface.[2] I climbed a prominent knoll about 400 meters south of the Sphinx and due west of a massive stone wall bearing the ancient name Wall of the Crow to take measurements and observe the pyramid site (see photo on page 152). The knoll is about 60 meters above sea level. The ridge that leads up to this point curves south and then southwest, west, and northwest, creating a natural sandy bowl. Lehner postulated that a workers' village might have been located here.

Lehner consulted old photographs showing that floodwaters reached areas close to the pyramid in the early 1900s. He also found maps showing old, unused irrigation channels, perhaps the remains of ancient canals constructed to bring water to the harbor. On the basis of this circumstantial evidence, he concluded that all of the features necessary for a huge construction camp were present.[3]

In his view, a harbor was established in front of the Sphinx and slightly to the southeast. The workers' village was in the bowl to the west, but a settlement of more permanent and finer houses was built east of the bowl, bounded on the north by the Wall of the Crow. Access roads connected the harbor and quarry with the construction site. Lehner elaborates on his concept of the Giza site plan in several of his publications.[4] Because large portions of the ancient site are buried under the modern village adjoining Giza, we may never know exactly what it looked like. But Lehner provides a convincing example of how the ancient Egyptians might have planned and developed the site to support the tremendous public works project that resulted in Khufu's pyramid.

Initially there was no infrastructure at Giza. Once the site had been selected, a small crew would have come and set up tents and other temporary shelter while preparing the site to receive construction materials—and a larger workforce. The initial tasks were to open up the quarry, level the site, build housing and workshops, establish a mess hall or commissary for preparing and serving food, and build access roads and a harbor to deliver people and material by ship. A canal connecting the harbor to the Nile River was excavated. Once the site of the pyramid had been pinned down, sites were selected for housing the workers and constructing various workshops close to the construction area. Workshops were required for masons, carpenters, weavers, artists, metalworkers, painters, toolmakers, and others. There also was a need for storehouses, kitchens, bakeries, a brewery, and quarters for the priests, officials, and administrators who oversaw the work. Other site needs included facilities for water supply, medical care, and sanitation. I know from my inspection of the site, where I saw the quarry and the remains of ramps and roads, and from the recent investigations of Hawass, Lehner, and others that all of these facilities must have existed.

During the early stages of the project, Hemiunu's project team developed a preliminary design, then a detailed design, and finally a detailed work plan. Recruitment of a workforce commenced. Then the first site work could start—surveying, building temporary housing for the workforce that built access roads, opening up the quarry, and grading the site.

CONDITIONS AT THE GIZA SITE

From the Nile floodplain, the land rises in the west to the plateau. The difference in elevation from the east side of the base of Khufu's pyramid to the valley floor is about 40 meters. The difference varies as the plateau slopes gradually toward the south. Access roads and eventually the causeway leading up from the valley temple to the mortuary temple, next to the pyramid, needed to be graded to establish the proper slope so materials could be moved easily up to the construction site.

The pyramid area, quarry, access roads, and other necessary infrastructure were no doubt surveyed. My estimate is that the grading of access roads from the quarry and harbor to the site required cut and fill of 20,000 cubic meters of sand, soil, and rock. Furthermore, the survey would have showed that the limestone fields extended south and east of the pyramid site for an area of 600 to 700 cubits by 800 cubits. From measurements made by excavation and at natural escarpments where the limestone face was exposed, the thickness of the limestone was determined to be 60 cubits and in some locations almost 80 cubits thick.[5] Assuming a typical block size of 2 by 2 by 3 cubits, the surveyed area easily could produce 2 million blocks, allowing for 10 percent for temples and other structures and 30 to 40 percent for trimming and waste, without making any allowance for the casing stones, which would come from the royal quarries at Turah. Clearly, limestone was present in sufficient quantity to build the pyramid. The main quarry would be located south of the selected site, and the stones would be transported up to the plateau by a sloping road leading directly from the quarry up to the site. Areas to the east were close enough to the construction site for a workers' village, shops, and storerooms. Everything was transported to the site and erected there.

A canal leading to the river was excavated, and a shallow-water port was constructed at the edge of the plateau for transporting stones from Turah and Aswan. In addition to the canal, which afforded access to the main channel of the river, a turning basin was required so vessels and barges could maneuver and discharge their cargoes. A stone wharf provided a means of unloading vessels.

A fleet of barges with a nominal displacement of 100 tons was needed. They would have a draft of 3 cubits, a beam of 8 cubits, and a length of 27 cubits. A vessel with this capacity could transport the large granite beams needed, for example, for the roof of the King's Chamber. They would be brought 900 kilometers downriver from Aswan to Giza. Smaller sailing vessels would also be used for bringing in supplies.

While Hemiunu's architects planned the port, the shipbuilders began working on assembling materials for the barges. A road leading from the port enabled materials and supplies to be delivered.

Nothing could be expected to grow in the desert, so provisions were made for supplying food. This required the construction of granaries, bakeries, and breweries. Livestock could be brought up to the site from the north and kept in pens until needed to feed the workforce. Initially, the heat and the dry desert winds made life difficult, but there was no alternative. The pharaoh's tomb was far too important to worry about inconveniences.

PLANNING THE WORK

Once the broad outlines of the project had been spelled out by Hemiunu, the planners went to work. The base of the pyramid itself would occupy an area of 193,600 square cubits, and the entire funerary complex, with the support facilities, the valley temple, the causeway, and the courtyards and walls, was easily ten times larger. Although the site was generally flat, a vast expanse of land had to be cleared. Wind-blown sand had to be cleared away and the exposed bedrock leveled. At the selected location, the footprint of the pyramid overlapped a limestone massif, which protruded into the northeast corner. The designers decided that it would be incorporated into the pyramid itself, anchoring the pyramid to the plateau. This planning decision reduced the leveling and site work and eliminated the need for thousands more stone blocks from the quarries.

We have Petrie to thank for an exhaustive study of the angle of inclination of the pyramid. He concluded that the as-built angle is 51°52 ± 2′.[6] The ancient Egyptians did not measure angles using degrees, but instead measured the horizontal setback of a surface from a 1-cubit-high vertical line: the seqed.[7] Several researchers have noted that a seqed of 5 palms, 2 digits is a close approximation of the measured inclination of Khufu's pyramid and therefore conclude that this was also the design slope. Expressed differently, since a cubit equals 28 digits, and since 5 palms, 2 digits equals 22 digits, the slope is 11:14—that is, the setback is 11 cubits at a height of 14 cubits. The arctangent of this angle is 51.84°, which falls within Petrie's range of error. It stands to reason, though, that no matter what method of angle measurement was used, it would have

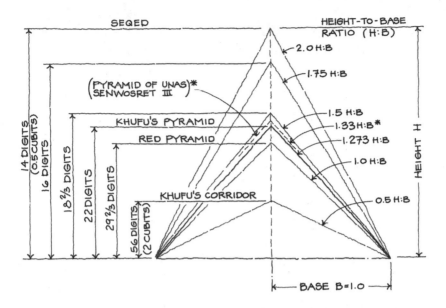

Determining the pyramids' angle of inclination.

been easy to apply in the field. It would not have used fractions. So how was the angle selected?

The slope of the pyramid—51.9°—is dictated by several considerations. First, to achieve the maximum height with the least material, the slope should be as steep as possible. However, the steeper the slope, the more difficult the construction and the greater the risk of structural damage, as occurred with the Bent Pyramid. An angle of 51.9° seems a reasonable compromise. In addition, given the ancient Egyptians' highly refined aesthetic sensibilities, one can assume they would have rejected a short, squat pyramid (base-to-height ratio 1:1) as well as something approaching 1:2. We know they were familiar with the 3:4:5 triangle because they used it to construct a right angle. The base-to-height ratio in this triangle is 1:1.33. Khufu's pyramid, at 1:1.27, comes close to this aesthetically ideal standard. There are some curious mathematical relationships at this angle. For one, the area of one face of the pyramid is *approximately* equal to the height squared. Also, the radius of a circle, the circumference of which is equal to the perimeter of the base of the pyramid, is *approximately* equal to the height of the pyramid. Did the ancient

Egyptians attach any significance to these quirks, or is it purely a mathematical coincidence? Most likely we will never know. Roger Herz-Fischer, a mathematician, carefully reviewed these and fifteen other theories concerning the pyramid dimensions and concluded that the design was most likely driven by aesthetic and constructability considerations, rather than by a theoretical method.[8]

WORK BREAKDOWN STRUCTURE

The incredible record of structures, tombs, and monuments left by the ancient Egyptians is convincing evidence of their engineering and construction skills—evidence that they had the ability to plan and execute complex projects. There is no doubt that they had a well-organized plan and executed it. No one knows the exact approach taken by Hemiunu to organize the work; few records remain, and those consist primarily of graffiti left by the various construction work crews who built the pyramid. But the very fact of distinct crews with specialized duties is evidence that the work was carefully organized.

Based on the evidence cited above and my own experience as a program manager, I developed a theoretical model of how an ancient Egyptian program management team would have proceeded. Hemiunu's first task was to identify all of the elements of work that needed to be done to determine the labor force, materials, and other resources needed. Under Hemiunu's leadership, each specialist in the program management team would have determined the various tasks that had to be performed. Unfortunately, no written record of this work has ever been found. We can be reasonably certain that it included the elements listed in the chart opposite. In the lexicon of a contemporary program manager, this would be called a "work breakdown structure."

Using the work breakdown structure, seven Level 1 overseers reported to Hemiunu, who had oversight of the entire project. Their responsibilities included administration and control of the project from a financial and records-keeping viewpoint, and real estate—identifying which of the royal lands would be dedicated to the construction of the project (and later to its permanent endowment for support of the king's

WORK BREAKDOWN STRUCTURE

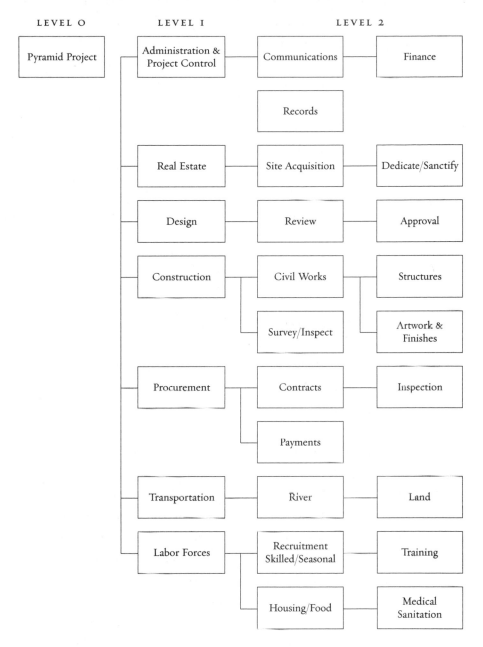

ba and ka for all of eternity). Responsibility for design fell to an overseer of architects, engineers, and others who would do the final design of the project. Supervision of construction probably rested with an overseer who was a veteran of a dozen other royal projects and had of late helped finish some of Khufu's father's last projects. The overseer of procurement reported both to Hemiunu and the royal treasurer. One overseer would have been responsible for shipping and transportation, and one for recruiting and supporting the massive labor force the project required.

Each of these Level 1 overseers in turn had a group of Level 2 supervisors responsible for subtasks. For example, under the overseer for construction, it is likely there was a supervisor for roads, the port, and all civil works; another for surveys, inspection, and control of dimensions; another for structures; and another for artwork and finishes. Additional overseers who functioned as quarry masters at Aswan, Turah, and Giza were responsible for ensuring the timely production and flow of stone to the project.

Project Organization and Construction Logic Diagram

Since it is well established that Fourth Dynasty Egyptians possessed excellent administrative skills, we can be certain that Hemiunu created a project organization, although only hints of how it was structured have been found (see Chapter 8). As a minimum, it had to cover the areas shown in the chart opposite.

Next, Hemiunu's program management team would have developed a "construction logic diagram" to illustrate the construction sequence that the various specialists found most plausible. The required steps would be easy to identify because these specialists were privy to the lessons gleaned from the pyramid construction efforts at Saqqara and Dahshur a few decades earlier. Supervisors and overseers made labor and material estimates for each of the tasks identified, and the team no doubt relied on the experience of its individual members, all of whom had worked on related projects. In addition, they had at their disposal the records and memories of the earlier pyramid builders and could use this information to perform reality checks on their own estimates.

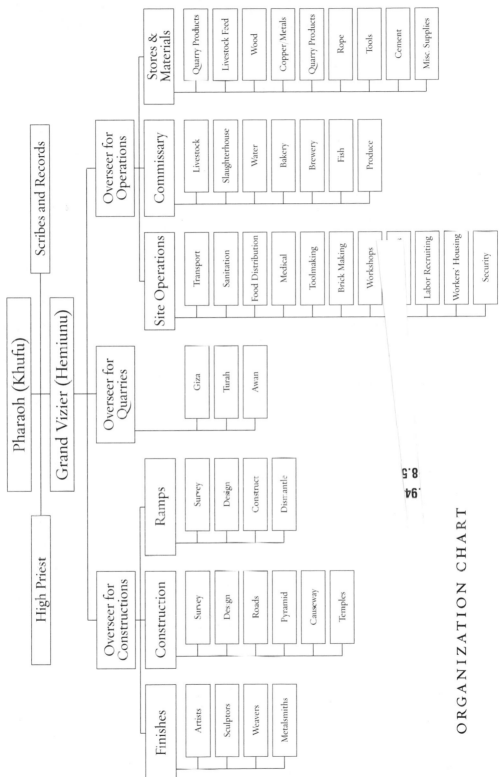

ORGANIZATION CHART

Pharaoh (Khufu)

High Priest

Scribes and Records

Grand Vizier (Hemiunu)

Overseer for Constructions

Overseer for Quarries

Overseer for Operations

Finishes
- Artists
- Sculptors
- Weavers
- Metalsmiths

Construction
- Survey
- Design
- Roads
- Pyramid
- Causeway
- Temples

Ramps
- Survey
- Design
- Construct
- Dismantle

Overseer for Quarries
- Giza
- Turah
- Awan

Site Operations
- Transport
- Sanitation
- Food Distribution
- Medical
- Toolmaking
- Brick Making
- Workshops
- Labor Recruiting
- Workers' Housing
- Security

Commissary
- Livestock
- Slaughterhouse
- Water
- Bakery
- Brewery
- Fish
- Produce

Stores & Materials
- Quarry Products
- Livestock Feed
- Wood
- Copper Metals
- Quarry Products
- Rope
- Tools
- Cement
- Misc. Supplies

Construction Logic Diagram

1. Mobilize project supervisory team
2. Prepare designs and plans
3. Site preparation at Giza
   Excavate to bedrock
   Rough grade
   Prepare access roads
   Prepare workers' village
   Prepare workshops
4. Shipping: acquire or build vessels
5. Mobilize: Giza quarry workforce
   Begin cutting blocks
6. Level base
7. Survey: pyramid base
   Establish working points
8. Start Descending Corridor and Lower Chamber
9. Prepare dock at Giza
10. Turah quarry
    Mobilize workforce
    Procure or fabricate tools
    Cut blocks
    Transport blocks to dock
    Load and ship blocks to Giza
11. Aswan quarry
    Mobilize
    Access road and dock
    Mobilize workforce
    Tools
    Cut granite for beams, lintels, sarcophagus, etc.
    Load and ship granite to Giza
12. Unload vessels at Giza
    Mobilize sleds
    Transport blocks to pyramid
13. Begin first course

14. Layout and mark: test-fit blocks
15. Build ramps
16. Start laying higher courses
17. Begin Ascending Corridor
18. Construct Queen's Chamber
19. Continue laying higher courses
20. Construct Grand Gallery and King's Chamber
21. Transport and set sarcophagus
22. Construct Relieving Chambers above King's Chamber
23. Set upper courses
24. Place capstone
25. Begin placing casing stones
26. Begin dismantling uppermost ramp
27. Place additional casing stones
28. Dismantle additional ramps
29. Continue placing casing stones
30. Dismantle remaining ramps
31. Construct courtyard
32. Construct mortuary temple
33. Construct causeway
34. Construct valley temple
35. Demobilize

Certain activities hinged upon others. For example, little could be done until a preliminary design of the pyramid had been completed, since it established all of the basic parameters. Once Khufu had approved the preliminary design (perhaps embodied in a stone model or an artist's sketch), then the site selection could be finalized, recruitment of the workforce could begin, and schedules for materials production from the quarries could be established. Although the site work could begin early, actual construction of the pyramid depended on several other activities. First, some of the civil works were essential—especially the access road from the quarry to the site. The boundaries and retaining walls of the main construction ramp had to be laid out so that as the pyramid rose, the ramp could rise with it. A small force of stonemasons could start early, cutting stones and preparing a stockpile in advance of the start of construction. But as the pace of the work picked up, more quarrymen had to be found or trained, toolmakers were recruited, and dozens of metalworkers were needed just to sharpen chisels for the crews in the quarries. The casing stones from Turah would not be needed for several years, so their production could follow a more leisurely schedule. Workers at Giza provided schedules of dimensions to Turah, and the white limestone casing stones were rough-cut, numbered, and stockpiled for later shipment to the site, where they would be dressed to their final dimensions and polished. At Aswan, the granite beams and panels for the King's Chamber, as well as the sarcophagus itself, had to be ordered long in advance because of the time it would take to produce and ship them downriver to the site. Work on granite members started early in the project, as soon as the planners had decided on the dimensions of the King's Chamber and other structures in which granite was employed.

CONSTRUCTION SCHEDULE

Unit time estimates (e.g., the number of labor-days per cubic meter required for excavation) were combined with supervisors' and foremen's estimates of materials to derive the amount of materials and labor required to perform each element of the work, as well as its duration. The data were used to develop a construction schedule that included the major milestones by which Hemiunu could measure the progress of the work and anticipate the resources that would be required a year or even two years into the future. He visualized the broad outlines of the project—as do most program managers. For example, after a thousand days, they would be ready to lay the first course of masonry. Another thousand days, and the King's Chamber would be done. Another thousand days, and the pyramid itself would be complete. (Although in Hemiunu's time it took weeks of work by a team of scribes to complete an analysis of the schedule, today detailed schedules for complex projects can be assembled in a few days using a fast digital computer. See Chapter 9.) Establishing a schedule was very important in ancient Egypt because religious consid-

First Thousand Days: Months 1 to 33

· Design
· Site preparation
· Labor recruitment
· Workers' village
· Surveying
· Initiate quarry operations
· Excavate site

Second Thousand Days: Months 34 to 66

· Descending Corridor/Lower Chamber
· Construct ramp
· Courses 1 through 55
· Ascending Corridor
· Middle chamber
· King's Chamber

Third Thousand Days: Months 67 to 100

· Elevate ramps
· Courses 55 through 218
· Place capstone
· Place casing stones
· Remove ramps
· Complete temples
· Clean up site

HEMIUNU'S MILESTONE SCHEDULE

erations dictated the favorable times for certain activities, and the work had to be coordinated with the flooding of the Nile to facilitate transportation. Hemiunu was able to draw upon historical experience gained in building the earlier pyramids. He knew that the Red Pyramid had been completed in ten years.[9] Above all, the king's tomb needed to be ready when he went to join the other gods in the western sky. It would not do to have it partially finished when the pharaoh's time arrived.

The analyses done to prepare the construction schedule enabled Hemiunu and the program management team to reject as impossible certain hypotheses related to construction methods. For example, a single ramp to the top of the pyramid would have been 846 meters long and involved more materials and construction work than the pyramid itself. Any ramp could not block access to the quarry and yet ideally needed to follow the shortest path from the quarry to the job site. Experience had shown that there was a limit to how steep a ramp could be. It had to be strong and stable, yet capable of being easily dismantled when the work was done. Finally, it had to be big enough that multiple teams of laborers could simultaneously deliver stones at the rate required by the schedule. Chapter 6 addresses how Hemiunu solved the ramp problem.

Hemiunu's schedule analysis showed that the production of blocks from the quarry was not a constraint. Stonecutting crews could keep up with the stonemasons who trimmed and placed the blocks. As an additional safeguard, it is likely that blocks were prepared in advance and stockpiled on-site to mitigate transportation problems or a shortage of quarry workers. Also, preproduction leveled the rate of production required during the early stages of the work. The program management team had to work out the details and logistics of site preparation, quarry operations, transporting the limestone casing stones from Turah and granite from Aswan, creating a workers' village for permanent skilled staff, constructing the ramps, performing the finish work, and removing the ramps at the completion of construction. Lists of materials were required. Messengers were dispatched to the royal quarries at Aswan, Turah, and other locations, where workers would begin the production of the sarcophagus, chamber roof beams, casing stones, and the other special structural elements.

With the preliminary master schedule in hand, Hemiunu could begin to assess labor requirements for the project. At the early stages, the primary needs were stonemasons and quarry workers, as well as laborers to excavate and grade the site and construct roads. When ramp construction got under way, more laborers would be needed, and still more when the major effort of hauling cut blocks of limestone up the ramps began. Someone was placed in charge of recruiting skilled workers and laborers. The recruitment plan took into account that once the crops were harvested, a large number of agricultural workers could be brought to the site on a temporary, seasonal basis. Thousands of unskilled laborers were ultimately required to assist with transporting the blocks up the ramps to the working area, where skilled masons would put them in place and build the corridors and chambers. With these tasks complete and all in readiness, the order was given to initiate the preconstruction work at the site. Crude shelters were erected, and the first materials arrived at the site.

Then it was time to consecrate the site. Khufu, accompanied by the high priest and others, symbolically marked the four corners of the site in a ritual known as Stretching the Cord. Evidence of this ceremony comes from fragments of a Fifth Dynasty relief in the sun temple of Niuserre and inscriptions found in several New Kingdom temples.[10] Although the ancient records describe the backdrop for the ceremony, we have to resort to imagination to fill in the details of the pomp and splendor. Khufu no doubt arrived in his royal palanquin, supported on the shoulders of muscular men who also served as his personal bodyguards. Like Khufu's mother's palanquin, his was constructed of fine Nubian ebony and bore several inscriptions, including his title of protector of the two kingdoms. He wore his double-crowned hat: the white miter with the head of a cobra of Upper Egypt, and Lower Egypt's red crown, symbolizing the union of the Upper and Lower Kingdoms. In his hands he carried the symbols of his office: his staff and flail. The high priest and several associate priests, dressed to symbolize Thoth and the goddess Seshat, the patroness of design and calculations, were in attendance, along with scribes, other functionaries of the court, and, of course, Hemiunu.

Khufu, in the company of the priests, examined a large stake that had been placed in a socket cut in the bedrock, marking the northeast corner of the pyramid courtyard. A short distance away, a corner of the rock massif was carefully cut to the form of the first three courses, symbolizing the actual pyramid itself. The rest of the huge site would have been cleared and leveled, ready for construction to begin.

Khufu next exchanged his staff and flail for a royal *bay* (see drawing on page 74). This instrument was inserted into a hole in the top of the stake, and by sighting through the aperture at the top, he could line up two small stakes in the distance, marked by red ribbons or some other marker. These marks had been placed by the survey team to indicate a true north–south line. Khufu observed that the survey stake was perfectly aligned to this reference bearing. This sighting by the king was pure ritual, in keeping with ancient traditions; the actual surveying had been done by others.

The surveyor's rod—10 cubits long—was as tall as three men. According to ritual, while two priests held it horizontally, Khufu verified that the scribed marks on it were each 1 cubit apart, as measured by a royal cubit. A priest or surveyor placed the surveyor's rod next to the marker at the northeast corner of the pyramid site and verified that it lined up perfectly with another marker 10 cubits south of the first. Forty-four such markers between the northeast and southeast corners proved that the foundation was of the prescribed size. Then, as the king walked around the perimeter of the site, an aide or priest unwound a cord tied to the corner stake. The process was repeated until the perimeter was marked.

When all was in order, again according to ritual, Khufu picked up a handful of Giza sand and tossed it into the enclosure, symbolizing the commencement of construction. From this point forward, the project had a name: Akhet Khufu (the Horizon of Khufu).

It is interesting to note how certain modern customs can be traced to ancient times. The politician who throws the first ceremonial shovelful of dirt at the groundbreaking ceremony for a new public building is reenacting an ancient Egyptian rite. Likewise, although today we may call it "program management," the methods of organizing large workforces and marshaling resources for public works projects were practiced by the Egyptians. The reader who wishes to learn more about the methods and tools of program management will find them in Appendix 4.

CHAPTER 5

THE DESERT
COMES ALIVE

The logistics of undertaking a huge public works project in the desert are daunting, as can be seen from reading the early history of the construction of Hoover Dam. At the end of the Great Depression, men were hungry for work. Construction workers flocked from the western states and even farther to sign up for a job at the project site. A tent city sprang up near what today is Boulder City, Nevada. Conditions were tough. Basic necessities were provided by the construction company, but initially there was no electricity, no running water. Temperatures in the daytime reached 45°C (110°F). It was not long before the first bar opened and prostitutes arrived to relieve the workmen of some of their earnings. Roads were gradually improved, utilities were constructed, and workers were joined by their families. A similar sequence of events occurred at Giza thousands of years ago.

During and following the Second World War, my firm, DMJMH+N, served as the designer, constructor, or program manager for the development of military bases and research facilities at numerous remote locations, from Antarctica to South Pacific islands to Saudi Arabia and the Sinai Desert. A more recent counterpart to these projects is the construction of the modern towns of Yanbu and Jubail in the vast, empty stretches of Saudi Arabia's deserts. Here an army of construction workers from dozens of nations and working for many international companies descended on the desert and created entirely new cities, with all of their infrastructure, in previously barren areas. Yanbu lies in the west, near the Red Sea; Jubail is on the opposite coast, on the Arabian Gulf. This work was led by two great contemporary program managers: at Jubail, the Bechtel Corporation, and at Yanbu, The Ralph M. Parsons Company.

The challenge at Giza was similar. There is no indication that Giza was occupied in any significant way before the Fourth Dynasty. We can reasonably assume that the first challenge facing the pyramid builders was logistical—building a construction workers' camp in the desert. Once this was done, the balance of the site work could be initiated.

SITE PREPARATION

The work needed to ready the site for pyramid construction was extensive. Roads connecting the construction site with the quarry and the harbor needed to be surveyed, staked out, and graded. Stone was gathered to surface the roads where the existing material was soft sand. At higher elevations, the road traversed bedrock limestone. Here, the material could be removed, crushed, and used to fill in and surface the low spots. Where heavy traffic was expected, the Egyptians cut paving stones and placed them on the roadway.

Nearly 800 meters of road were required. I considered two main roads. The first ran from the quarry to a lay-down area near the end of the main ramp, a distance of 400 meters with a rise of 30 meters and a width of 50 meters. I assumed a balanced cut-and-fill operation requiring removal to an average depth of 1 meter over one-half the length of the road. My second road ran from the harbor to the lay-down area. I assumed this road, which rose 45 meters from the elevation of the harbor, was 400 meters long and 25 meters wide. This I also assumed was a balanced cut-and-fill operation, with a cut to a depth of 2 meters over one-half the length. Constructing these roads would require the excavation and transport of 20,000 cubic meters of rock and soil, and the placement (as fill) of a like quantity.

Next, the temporary housing area was expanded into a permanent workers' village to house the core group of supervisors, stonemasons, artisans, and skilled tradesmen responsible for the construction program. Fairly intact workers' villages have been excavated at three sites: Kahun, Tell el-Amarna, and Deir el-Medina, dating from the Twelfth to the Eighteenth Dynasties.[1] The similarities in the design of these workers' towns suggest that a typical plan was followed. We can assume that the workers' village at Giza was similar.

The archaeological evidence indicates that the houses were dark and small. Furniture was limited to storage chests, stools, a table, headrests for sleeping, and one or two other items. Baskets and jars were used for storage. Raised benches of brick or stone were used for couches and beds. Larger houses were made available to overseers and senior craftsmen. There is some evidence that the state provided housing, or at least temporary housing, but in general workers were largely responsible for building their own houses with mud bricks furnished by the state.

The houses at Deir el-Medina were single-story, flat-roofed structures. The village had seventy houses within a walled area 132 meters long by 50 meters wide. Later, as the community grew, additional houses were built outside the walled city. Doors were made of wood, and the stone doorjambs or lintels were painted with red hieroglyphics listing the names of the occupants. The houses were laid out in a regular, rectangular pattern, the entrance facing the street.[2]

In the living room was a couch or seat fabricated from bricks. Niches in the walls contained statues of local gods. The sleeping room had low brick platforms for beds. The kitchen was generally an open area with a brick or pottery oven for cooking. There were facilities for storing food and grinding grain.

To estimate the labor required to build the workers' village at Giza, I assumed there were four types of houses. For foremen and supervisors, I assumed a lot 6 meters wide by 10 meters long with a mud-brick structure 4 by 10 meters divided into three or four rooms: the outer hall and entrance, the living area, a bedroom, and a room for cooking and storage. Each house had an attached courtyard that could also contain a small garden. The houses were roofed with palm thatch or reed mats supported on palm logs. Floors were hard-packed earth. The more affluent workers would have plastered and perhaps painted the interior walls. Windows were small openings high on the walls. Door openings were covered with woven mats or in some cases wooden doors.

Skilled tradesmen with dependents had similar but smaller houses (4 by 7 meters) with three rooms on a 42-square-meter lot. Workers without dependents had even smaller houses (3 by 6 meters) on 24-

square-meter lots. In addition, there were several villas (200 square meters) for overseers. With allowances for a commissary, bakeries, workshops, and breweries, as well as narrow streets dividing the village into neighborhoods, around 3,500 dwelling units could be accommodated in an area of 150,000 square meters (390 meters square). With these assumptions, I estimated that the construction of the workers' village required more than 8 million mud bricks to reach full occupancy.

WORKFORCE

Recent excavations at Giza by Hawass and Lehner are now revealing previously unknown details concerning the site infrastructure and the workers who built the pyramids. Lehner is systematically uncovering a large complex of structures southeast of the pyramid near the Wall of the Crow. He and his team have found evidence of breweries, bakeries, a fish-processing facility, a copper-tool shop, storerooms, and other structures.[3] (See Plate 17.)

Elsewhere on the site—south of the quarry area—the remains of stone kilns can be seen. Pieces of characteristically green copper rust (copper carbonate) indicate that copper smelting and the manufacture of copper tools for the stonemasons was done here.[4]

On the hillsides overlooking Lehner's dig, Zahi Hawass is excavating the tombs of the workers who constructed the pyramids. From inscriptions within the tombs, statues and carvings, and tools and other household objects found in these tombs, he has been able to reconstruct a fascinating picture of the daily lives of the workers.[5]

His work reveals that entire families lived at the site. Some tombs contain three

Excavating the workers' village.

generations. A father passed on the skills of his trade to his son, and he to his son. The inscriptions reveal that the workers were artisans, craftsmen, supervisors, foremen—and that all took great pride in their work. They were there willingly—they certainly were not slaves—and apparently highly motivated by what they perceived as a project of great importance for the kingdom. The tombs give some indication of how the various trades—carpenters, masons, painters, sculptors—were organized into teams with supervisors, and these into larger groups, and so on. The organization of the workforce is described in more detail in Chapter 8.

These workers were paid for their services in the necessities of life: food, including grain, cooking oil, bread, beer, fish, and other foodstuffs; and cloth and other needed materials. Pay varied according to rank. Foremen received more than craftsmen and laborers, supervisors still more, and so on.

These men worked hard and died young. Hawass reports that medical examinations performed on a number of the exhumed mummies of workers show deformed vertebrae, broken bones, and other signs of accidents and hard labor. They also display evidence of medical treatment, surgery, bones that have been set—even skull trepanning. It took a rugged individual to undergo treatments of this sort without anesthesia, unless, of course, the ancient Egyptians had developed some long-since lost method of blocking pain.

The evidence is overwhelming that these were not slaves. Slaves would not have been paid and afforded the obvious respect these men received, nor would they have received medical treatment. Slaves would not have been honored by burial in elaborate tombs near the site of the pharaoh's tomb. More likely, the bodies of slaves would have been cast into the emptiness of the Western Desert and left for scavengers.

Based on the size of the workers' cemetery, Hawass has estimated that the permanent workforce at Giza—the central core of skilled masons, craftsmen, and artisans—numbered 5,000.[6] Lehner estimates as many as 4,000 core workers, with 20,000 to 25,000 permanent residents at the site, based on the size and apparent capacity of the ruins he has uncovered.[7] Excavations of Deir el-Medina (occupied during a much

A general view of the workers' cemetery.
INSET: Some individual tombs in the workers' cemetery.

later period) similarly support the theory of a permanent labor encampment at the site. Extensive records obtained there document many day-to-day details about the workforce, including organization, pay rates, the settling of family disputes, and the first organized strike on record, during which workers who had not been paid refused to work.[8]

SITE FACILITIES, TOOLS, AND CONSTRUCTION METHODS

Behind Khafre's pyramid are the remains of long, narrow rooms. It was once believed that they were workers' barracks. More recent evidence indicates that they were workshops and possibly storage facilities for tools and materials.

No such workshops have been found at Khufu's pyramid. There can be little doubt, however, that they once existed. Workshops were required for carpenters to make sledges, scaffolding, levers, rollers, and other forms of construction lumber, as well as the limited furniture in the workers' housing. Other workshops were required for metalsmithing, rope making, textile production, sculpting, and other trades.

To estimate the labor and materials required to construct workshops, I assumed seventy-five workshops, similar to the "galleries" at Khafre's pyramid, encompassing an area of 7,000 square meters. For breweries, bakeries, butchers, and kitchens, I assumed an area of 7,500 square meters, based on the food preparation requirements for a permanent on-site workforce of 5,000 people. Once the permanent workforce was well established, food would have been prepared in their homes. Also, a transient labor force throughout the duration of the project made on-site food preparation a necessity.

To produce mud bricks used in constructing houses, workshops, and facilities for food processing and preparation, I decided that the planners placed four brick factories at convenient locations near the construction sites and the source of raw materials to minimize transportation labor. Mud-brick production does not require much in terms of a building. It is primarily an open-air operation with an area for mixing the raw materials (usually a pit in the ground), another area for casting the bricks in molds, and a larger area for drying the bricks in the sun. A small shop would be required for storing tools and brick molds, repairing brick molds, and keeping supplies and records. Each brick-making operation took about 400 square meters. Combined brick production was 5,000 bricks per day, or 1.5 million bricks per year.

The ancient Egyptians made rope from flax, rush, papyrus, palm fibers, esparto grass, and halfa grass.[9] Ropes were made by twisting fibers into individual strands and then making ropes of from three to five strands. Tomb paintings show men standing and twisting together two strands. One end was fastened to a post while the other was twisted. The rope was moistened and pounded to compress the fibers and strengthen it.[10] Ropes have been found with diameters of 6 to 7 centimeters.

The high-quality cordage made in Egypt was exported to other countries along the Mediterranean shore and used for oceangoing sailing vessels.[11] Excellent examples of Fourth Dynasty ropes were found in Khufu's boat pit. Working strengths of up to 5 metric tons may have been feasible. Such strength required a large-diameter rope that would have been difficult for laborers to pull without attachments or handling loops of some sort.

To estimate the working tensile strengths of practical ropes, I used data on manila rope from the 1800s, took 50 percent of the breaking strength, and then applied a safety factor of seven to determine the safe working capacity:

Tensile Strength of Ropes

ROPE DIAMETER (CM)	SAFE WORKING LOAD (NEWTONS)
1.3	580
2.5	2,220
4.2	5,780
6.0	11,100

For heavier loads, larger ropes could be made, or multiple lengths of smaller ropes could be used. For example, if the above estimates are correct, five lengths of 6-centimeter rope could be combined to give a 5-metric-ton working load capacity.

I estimated that 120,000 meters of rope of various sizes were required for the project, assuming some rope was reused.

From early Predynastic times, the Egyptians developed and used copper tools and implements (see Chapter 2). Over the centuries, copper ores were mined locally in Egypt, in the Sinai Desert, and as far away as the Negev Desert at Wadi Timna.[12] The mines in the southeastern corner of the Sinai Desert (Wadi Maghara and Wadi Kharit) bear inscriptions referring to Djoser, Sneferu, and Khufu. Today, at least a dozen sites in the Sinai are known to have copper ore.[13] The use of copper for a wide variety of everyday objects in addition to tools, from drain-pipes to hand mirrors, suggests that it was plentiful and the technology for working it was well established.

Bronze (an alloy of copper with 4 to 10 percent tin) did not come into widespread use until after the Fourth Dynasty. Most copper tools from Khufu's time were made from copper with various additives, which may have been incidental or accidental.

Copper metal is derived from one of three sources: native, or free copper; sulfide ores, including minerals such as chalcocite (Cu_2S) or chalcopyrite ($CuFeS_2$); or oxidized ores such as malachite, azurite, cuprite (Cu_2O), and melaconite (CuO).[14]

Primitive smelters such as those found in Egypt and in the Negev Desert were basically pits in the ground. The ore—mixed with charcoal and other materials that served as a flux—was heated to the melting point by the charcoal, which served as both fuel and the reducing agent.[15] When the ore melted, the flux reacted with the impurities to form a slag, which floated on top of the denser molten metal.

The metal can be extracted easily from native copper by breaking up the ore and cleaning it to remove obvious impurities. The ore is then heated in a smelter in the presence of limestone as a flux. For oxide ores (typical of the ancient mines) the flux was probably a mineral such as hematite, Fe_2O_3, but the process was similar. For sulfite ores, the process is more complicated, since the ore must first be roasted (brought to a red heat in excess air) to remove volatile impurities and oxidize the metal sulfides.

Pure copper is a relatively soft metal. However, such impurities as arsenic harden it and are beneficial for producing tools. The presence of some arsenic is expected, given the ores used by the ancient Egyptians, but the extent to which the smelters knew of this benefit and were able to control the mixture is not known.

Numerous copper tools were found in the workers' village at Kahun. In addition to the tools themselves, molds for such cast implements as copper axe heads and knives were found. Copper was smelted locally, but the ores may have been brought from elsewhere.

A number of copper samples from Kahun were analyzed by British investigators who employed neutron activation analysis, a technique for accurately determining trace concentrations of metal impurities. Both tools and ores were tested.[16] All of the tools contained arsenic in concentrations higher than the ores themselves. The results are consistent with the historic evolution of copper tools: the early ones were made from soft native copper; later, in the Third and Fourth Dynasties, copper with arsenic appeared, and in the New Kingdom, bronze.

The copper tests showed that even as early as Kahun, the samples contained tin. In low concentrations, tin could be considered an impurity, but at higher concentrations, it has to be considered an intentional addition. One can see how a smelter, experimenting with various admixtures, could stumble on the invention of bronze.

Further evidence of copper production and copper tool availability has been provided by a team of archaeologists from the University of California at San Diego, working at a site in southern Jordan, about 50 kilometers south of the Dead Sea (as accessible to the Egyptians at Giza as the southeastern Sinai).[17] This extensive site, consisting of approximately seventy rooms, workshops, and courtyards, was apparently destroyed in an earthquake around 2700 BC. As a result of the earthquake, the researchers found hundreds of copper tools (chisels and axes), more than a thousand ceramic molds for axes, chisels, pins, and bars, and thousands of tons of slag, indicating that the facility produced hundreds of tons of copper during its lifetime. Much of the metal was cast in the form of ingots, and tracer studies show that some of this metal ended up in Egypt.

Given that the Egyptians knew how to make copper tools even before the First Dynasty, it would seem that by the Fourth Dynasty, 400 years later, this technology would be well advanced. In the Fourth Dynasty, tools were made by hammering (hot or cold) or casting. Cast items were further shaped by hammering, but care had to be taken not to make the material brittle. Saws, drills, and other cutting tools were used with an abrasive such as quartz sand. As grains became embedded in the surface of the softer metal, it would create an almost diamond-hard cutting blade.[18]

I made crude copper chisels with soft copper pipe and found it was possible to cut limestone with them, but the tools dulled rapidly and needed to be resharpened frequently. Khufu's stonemasons used heavy stone hammers and adzes for rough cutting and trimming, as well as copper chisels. I estimated that 3,500 stonemasons working 1,200 days would need between 300,000 and 500,000 chisels during the life of the job, assuming that each chisel could be resharpened 100 times before it had to be melted and recast.

The hearth smelters found in ancient Egypt are typically 30 centimeters in diameter and could have produced an ingot weighing 60 kilograms. Since a chisel weighs about 0.23 kilograms, one batch could produce more than 250 chisels. Three such smelters working for a year or two could produce all of the copper needed for chisels. If the copper were imported as ingots, 125 tons—several boatloads or a few dozen caravan loads—would suffice to make the chisels required for the project. This was well within the capabilities of the ancient Egyptians.

Wood for sledges, boats, and tools was another critical construction material. Egypt did not have native sources of good timber, so wood was imported from Lebanon and Syria. Byblos was only 600 kilometers away by sea, and a sturdy vessel could make the trip in four or five days. Rafts brought Lebanese cedar, cypress, fir, and pine wood to Egypt by sea, and Nubian ebony and African hardwoods were brought north down the Nile River.

The Egyptian carpenters also made effective use of local woods—sycamore, fig, acacia, tamarisk, sidder, and willow. These woods yielded small pieces of lumber that had to be joined together to produce storage chests, cosmetic boxes, stools, tables, coffins, carrying chairs, sledges, doors, tools, and other objects.

Egyptian carpenters had many of the tools we use today: squares, measuring rulers, saws, drills, hammers, adzes, and scrapers (see page 83). Sand and sandstone were used as abrasives. They had glues and paints. Joints were usually pegged, although copper nails were used later. Some wood objects were decorated by inlaying them with ivory, faience, or semiprecious stones, or overlaying them with copper or gold sheet.[19]

A major allotment of wood was used for sledges during construction of Khufu's pyramid. I estimate that 1,200 sledges in at least four different sizes were required. Hundreds of levers or pry bars, 1,000 rollers, and miscellaneous lumber for scaffolds, shoring, and other uses were also needed. These items alone required around 400 cubic meters of lumber.

I assumed that logs 10 meters long and 50 centimeters in diameter could be imported and ripped into boards of varying sizes to meet the carpenters' needs. Such a log contains 2 cubic meters of wood. Allowing for 50 percent waste during cutting and trimming, 400 logs would meet the basic needs, without considering the additional wood available from local sources, which I assumed was sufficient to build houses and furniture for the workers' village.

The idea of a harbor near the site is logical, considering how Fourth Dynasty Egyptians operated. There are clear records of ancient ports at Aswan and other locations along the Nile. Aswan was important because it was the source of mammoth granite objects—statues, beams for construction, even obelisks. These were painstakingly cut from granite boulders or outcroppings, moved to the river's edge, and loaded on barges for shipment downriver.

Lehner points out that in ancient times, the course of the Nile was farther west.[20] During the Nile's annual inundation (July through October), water reached to within 400 meters of the site.[21] The harbor needed to be ready by July 2548 BC, when the Nile flooded, to bring in goods by ship. Otherwise the project would be delayed another year.

From the annual flooding of the Nile, Hemiunu's surveyors gained a good idea of the extent to which the Nile floodwaters spread across the valley and approached the margin of the site. The workers' village was above the high waterline. I believe the construction crews would have built a harbor at the closest approach of low water by excavating a shallow-water port, 10 cubits deep, with a stone seawall and quay for offloading barges and other vessels. I assumed that the quay was 50 meters long and 10 meters wide to allow for docking several barges at once and handling cargo. These dimensions are similar to those of the original

A Nile River canal.

quayside at the temple of Karnak, which had a sloping ramp or slipway for pulling cargo or small boats out of the water.[22]

I assumed that the harbor was circular with a radius of 75 meters—enough to turn vessels and anchor or moor those waiting to discharge cargo. A canal connecting the harbor to the river might have been 400 meters long, 12 meters wide, and 2 meters deep, which would accommodate vessels with a beam of 9 to 10 meters drawing up to 1.5 meters. The Nile canal shown above is similar to what might have been built at Giza. Embankments were constructed along each side of the canal so crews could maneuver barges when there was no wind. I also assumed that an impromptu shipyard was established nearby so the boatbuilders could begin constructing barges to carry stone across the river from Turah when the Nile flooded.

Siting the Pyramid

Several tasks were completed before Khufu performed the ritual of Stretching the Cord to signify his approval of the site and initiate construction. The surveyors laid out the rough boundaries of the areas to be leveled for the base of the pyramid. On the north side, some of the bedrock had to be cut away to level an area for the courtyard that would surround the pyramid. Once the site had been cleared and leveled, and a north-south baseline established, they had to cut and smooth the sides of the protruding bedrock in a stair step fashion to the same dimensions as the backing stones (1.5 meters high). The lower portion of the rock protrusion was trimmed to leave a smooth vertical face that lined up precisely with the row of heavy backing stones being produced in the quarry. The lower edge of the protrusion was cut to a height of 1 cubit, identical to the height of the platform stones that were laid down to do the fine leveling of the pyramid base (see photo on page 69). The site of the larger, permanent workers' village would have been laid out at this time as well.

In conjunction with grading the site, the location of the pyramid was established by selecting a point for one corner (say the northeast corner). Then a square measuring 230 meters on a side was laid out. The photo of a cornerstone at the Bent Pyramid below shows that the masons carved a 90° angle to mark the corner. The base was positioned to leave space for the courtyards and other structures that were constructed later.

The northeast corner seems an obvious starting point for Khufu's pyramid, since this is where the bedrock remained. According to Hawass, in this area the bedrock rises to a height of at least 17 meters in the pyramid.[23] The full extent of the bedrock outcropping that was incorporated into the pyramid is not known; only a portion of

A corner platform stone with orientation marks, Bent Pyramid.

A corner of the Red Pyramid (note size of blocks).

it can be seen at the northeast corner.

From a benchmark adjacent to the northeast corner, a measurement was made to mark the southeast corner benchmark, say at 240 meters. (This would place the benchmark 10 cubits [5.2 meters] away from the actual corners to avoid interference when maneuvering the blocks into position.) Using this radius as a starting point, the surveyors would then "back shoot" a true north-south line to the northeast corner, using one of several methods discussed earlier. Correcting for the measured offset, the true position of the southeast corner was then established.

Lehner has an excellent description of the process used by the builders to set the boundaries of the pyramid base.[24] He postulates that the surveyors installed a series of posts or stakes along the pyramid baseline.

These posts were placed in holes cut in the bedrock at regular intervals of approximately 7 cubits along the side of the pyramid. After the pyramid was completed, the holes were plugged with small stone plugs and covered by the courtyard paving stones. We are able to see them now because most of the paving stones have been stolen over the years.

A string of surveyor's postholes, Khufu's pyramid.

A hauling basket used by present-day workers.

LEVELING THE SITE

To level the site, the area would have to be excavated to bedrock. Loose sand and soil were removed and stockpiled at the site for later use. Once the underlying bedrock was exposed, leveling could begin. This was necessary because the rock formation slopes gradually to the southeast.

Fourth Dynasty laborers used reed baskets that contained approximately 0.25 cubic meters of material for hauling sand and soil, the amount a worker could carry and still put in a full day's work. Even today, vessels resembling these reed baskets are used by Egyptian workers. It is likely that the overburden of loose material was not very thick, but sand and eroded limestone covered the site. Even a few centimeters of material over the 90,000-square-meter area that was rough-graded amounted to a substantial volume of material to remove.

Once the bedrock was exposed, it could be cut down to the current finished grade on the north side, where the workers laid out a grid pattern on the bedrock and cut a series of channels in the rock. Once the channels were cut, levers were used to break off blocks of stone.[25] These blocks were removed and stockpiled for use elsewhere. After the approximate finish grade was reached, the grid marks were removed, and the surface was smoothed to receive paving stones. This last step was never done for the area shown in the photo.

The north side of Khafre's pyramid, showing method of leveling courtyard.

The site of Khufu's pyramid is flatter than Khafre's site. The approximate dip of the Mokkatam limestone formation from the southerly edge of Khufu's pyramid toward the Sphinx is 8°. I estimated that a cut-and-fill operation based on a height differential of 2 meters was required, which involved cutting and moving 37,500 cubic meters in a balanced operation.

Leveling was done in accordance with a series of benchmarks established on the site. The ancient Egyptians' leveling work was excellent. Edwards reports that the site is level to within an average deviation of ± 2 centimeters.[26] A modern leveling achieves high-end accuracy of about 0.3 centimeters and an average accuracy of 3 centimeters. The ancient Egyptians did very well, given their limited tools.

It may be that only the perimeter of the base area was leveled accurately, because it was probably not necessary to level the entire surface. In any case, the whole area would have been cleared to bedrock and cut flat to provide a solid, stable bearing surface.

To estimate the labor required for grading, I took the site area to be cleared as 300 meters by 300 meters (90,000 square meters). I assumed that one-third of this area required rock removal for an average depth of 2 meters (primarily on the north side). Sand and wind-blown debris had to be cleared over the entire area (except where the rock massif stood) to a depth of 0.5 meters. For the rock massif, I assumed that 10 percent of its volume had to be cut in a stair step manner to line up with the platform and backing stones:

Clear and grub: 33,750 cubic meters

Remove stone: 60,000 cubic meters

Shape massif: 22,500 cubic meters

Once the site had been leveled and graded, work could begin on the placement of the platform stones.

Quarrying

While the site was being graded, quarry operations commenced at Giza. Arnold provides an excellent description of ancient Egyptian stonecutting techniques.[27] Excavations were made in the quarry to expose the limestone, and vertical faces were trimmed. Lines were marked on the surface and the face with red ochre paint to guide the stonemasons. Channels were cut in the stone on three sides of the block and notches cut at the bottom edge. Then, using mallets, wedges, and levers, a rough-cut block of stone was broken loose and removed from the vertical face. A similar process occurred at Turah, where the royal quarry was swarming with workers who were marking new sections of the quarry to prepare for mining the white limestone casing stones for the pyramid.

I visited the Turah area in April 2000. Even today, forty-five centuries later, this area is famous for its stone quarries. The quarries are

The old quarry area, showing the limestone formation.

Partially cut stone in the old quarry.

hundreds of meters deep, and the stone is mined with power tools, cranes, and trucks instead of copper tools. With the kind assistance of several local businesses, such as the Cairo Marble and Granite Company, I was given a behind-the-scenes tour of the mining and stonecutting operations currently practiced. Electric power saws cut stone into blocks and slabs. Trucks left the plant loaded with blocks of white limestone weighing 10 to 12 tons—much easier than moving them by hand on wooden sledges.

Quarry operations also started at Aswan, an important source of granite during the Fourth Dynasty. Workers first laid out the profiles of the granite beams needed as major structural members in the pyramid, in accordance with instructions they had received from Hemiunu's designers. Then the laborious process of cutting out the granite began. Using round stones of dolerite, they chipped away the granite a few flakes at a time to make a groove along the line marked by the stonemasons. After much hard work, the groove became a channel, and then the channel became a slot deep and wide enough for a man to stand in sideways, until the stone had been cut to the finished depth. Next, the beam was broken loose and moved to another site, where it would be trimmed and polished to its final shape before being shipped downriver to Giza.

The largest operation, of course, was at the Giza quarry. Imagine what it would have been like to visit the quarry when it was in full production, with thousands of quarry workers and laborers cutting stones and moving them to the job site! One can almost feel the impact of a thousand hammer blows on stone, hear the chanting of the laborers pulling their heavy loads and see the workers in their rough tunics, brown with dust from the quarry. Each team of stonemasons works on an assigned block, and they systematically move down the face in one direction, removing blocks of stone as they go. Nearby, stacks of poles have

Turah limestone quarries today.

been fashioned into large pry bars. Sections of tree branches, used as rollers, have also been stored near the workers. Farther down the face, a large group of men surrounds one of the blocks, while the foreman for this gang of stonemasons stands on the quarry face above them, directing the work.

A block has been cut from the limestone face on three sides. A channel, about two hands wide, separates it from the solid rock on the left and right sides and on the back. All around the base, the block has been scored with a narrow groove to define a break plane. On the bottom front edge, the masons have chiseled out five triangular indentations (see photo on page 144). Ten masons—five holding wedges, five hammering—are working in unison to drive wedges into these indentations so that the force is applied uniformly across the front edge. Meanwhile, a score of laborers stand ready with long poles. A muted cracking sound and a slight shudder of the block signals it has broken loose, and laborers swarm around it with long poles, ready to lever it out of the quarry face. Others position rollers in front of it. With tremendous exertion by more than twenty men on ten levers, the block slowly moves forward a millimeter at a time. Gradually the massive block is positioned on the rollers. Once it is clear of the quarry face, some of the laborers store their levers and return with a heavy rope harness. Slowly the block is moved away from the work area to a staging area, where trimming is completed and the block is numbered with black ink. As soon as that block is clear, the stonemasons on the adjacent block resume their tedious job of chipping out the channels.

In this manner, the quarry workers prepared block after block for delivery to the laydown area at the base of the pyramid. From there, teams of stone haulers moved them up the ramp to the working course, where stonemasons put them in their final position.

Construction Begins

Once the site was graded, rough-leveled, and surveyed, a foundation platform was placed. It consisted of a series of flat stones, one cubit thick, that formed the first course of masonry (see photo on page 69). The fine leveling took place at this time. The platform stones were dressed as required to compensate for any irregularities in the leveling of the bedrock. It may not have been necessary to place the platform stones over the entire area, but as a minimum, they were placed along the entire perimeter and accurately leveled, as shown by modern surveys. With the level platform in place, the construction of the pyramid itself could commence.

Using the leveling benchmarks as a guide, the base area was dressed to provide a horizontal working surface, although possibly not all at the same elevation. In other sites, it can be seen that the workers stepped the foundation to minimize cutting.

In tandem with leveling the site, construction of the Descending Corridor began. This inclined shaft was cut to a depth of 30 meters below the surface, where a short horizontal segment was constructed. Then the Lower Chamber was cut into solid bedrock. Another horizontal tunnel was started in the south wall of the Lower Chamber. It extends for about 16 meters and then stops. Its purpose is unknown.

While the tunneling was going on, it is likely that a construction gap was left open in the pyramid base to facilitate the removal of debris. Once the tunnel was complete, the upper portion was framed with stone as the pyramid rose.

It is generally thought that the Lower Chamber was started and then abandoned.[28] However, Lehner has suggested that it was actually constructed last. While it was under construction, the pharaoh died, and the construction stopped and was not resumed.[29] However, since the upper portion of the Descending Corridor is framed with masonry and not cut through the existing layers, it seems that the original plan provided for a lower chamber.

The construction of each course was based on three components: the outer casing stones of carefully dressed white Turah limestone; an inner layer of backing stones, which aided structural integrity; and the core blocks of Giza limestone, which were not dressed accurately but fitted expeditiously into the inner volume of the pyramid and leveled only at the next step (not at every course). Irregular shapes were incorporated into the structure to maximize the use of available materials. Voids were filled with mortar, rubble, or debris.

Lehner provides a well-thought-out and plausible construction sequence for placing the foundation slab and the lower course of masonry.[30] He postulates that construction began at the northeast corner, where the bedrock projects into the pyramid. From this point the reference lines for the sides and other corners were established as described above.

Construction proceeded in layers above the base, until the next step, or reinforcing layer, occurred. At this point the structure would be leveled very carefully again. It was not necessary to level each layer, which would have increased the amount of cutting and trimming on each block and wasted material. In Khufu's pyramid, there appear to be fifteen to twenty courses per step and approximately twenty steps where larger backing stones are placed.

As noted previously, the thickness of the backing stones at grade is 1.5 meters. The thickness decreases with elevation fairly uniformly over the first fifteen courses, and then increases again. At Courses 1, 7, 35, 36, 44, 90, and 98, there are blocks 1 meter thick or more. At various levels there are double blocks, which overlap two courses and have thickness of 1.3 to 1.5 meters.

Recent studies of the pyramid using microgravimetric measurement techniques suggested that the pyramid has areas of reduced density. A French architect, Jean-Pierre Houdin, has advanced a theory that these areas mark internal corridors or ramps used in construction.[31] Another possible explanation is that the interior is not solid stone, but instead has a matrix of carefully placed stones providing structural stability, interspersed with voids filled with sand and rubble. This method of construction is observed in later pyramids and certainly would have reduced the labor required to build Khufu's pyramid.

The use of smaller blocks (say, 0.50 meters thick and 0.65 by 0.75 meters) reduces the difficulty of placement considerably. In this case, a smaller team of workers could push and pull the smaller blocks into position at the higher elevations of the pyramid, whereas the large backing stones clearly were skidded and then levered into their final position on the lower courses. A thin layer of liquid mortar was probably used as a lubricant to facilitate moving the heavy stones.

Construction continued up to Course 23, the elevation at the floor of the Queen's Chamber. As work proceeded, the Ascending Corridor was framed in. Meanwhile, the main ramp had to be gradually raised to remain even with the ongoing construction.

CHAPTER 6

SOARING TOWARD
THE HEAVENS

Were we able to visit Giza in 2547 BC, we would see the construction site swarming with thousands of workers, so many they could populate a small city. They are confident and proud, and with good reason. They are engaged in an unprecedented construction project, and they are executing it masterfully. They are building by hand—by their hands—a structural colossus that will endure for millennia as one of humankind's most brilliant achievements. If anyone could build a structure on this scale, they agree it would be Hemiunu, a man whose brilliance and vision are breathtaking. It surprises no one that he has conceived this extraordinary pyramid—and that his confidence in the plan he has devised for its construction is unflinching.

All involved in the effort are well aware of its magnitude, as they are of the scrutiny of Pharaoh Khufu, who monitors every aspect from the White Wall, his palace in the royal compound near Memphis. Hemiunu's orchestration of the work is superb. Swiftly and assuredly he has contrived an elaborate construction program that is unfolding with precision, marshaled an enormous army of expert workers and overseers, amassed vast inventories of materials from all over the Egyptian kingdom, and sectioned the Giza area into well-delineated sectors, including a sizable residential enclave. The large village in which the permanent workers live with their families supports the construction site, and residence in the village is viewed as prestigious by the stonecutters, stonemasons, carpenters, metalworkers, toolmakers, and various other laborers, overseers, craftsmen, and draftsmen who have vied to join this workforce. The best merchants in the kingdom are here, and theirs is a tight-knit, supportive community.

Those who come here—administrators assessing the progress or newly arriving residents—are struck by how many lives are focused on this construction effort. Once unoccupied desert, the Giza Plateau has sprung to life, its daily rhythm an assimilation of the workers' occupational and domestic lives. Even the smell of the place is an amalgam of construction site and village. The acrid fumes rising from the foundries blend with the pungent smoke of midday cooking fires. The aroma of baking bread wafts up from the village to mingle with the odor of human sweat. The plateau is never silent, never still. By day, it is alive with the clamor of construction: stonecutters chiseling the immense blocks of limestone; supervisors shouting instructions; workers calling to each other; teams of laborers dragging their heavy burdens up the ramps, chanting in a deep euphony that soothes the strain of their exertion. By night, it mellows into the hum of village life: families sharing evening meals; children pressing for extra minutes of play; men recounting the day's work and planning the next—sounds punctuated perhaps by the harsh staccato of a quarrel, the soothing strains of a lullaby. Even the early morning hours are not soundless. Muffled voices, muted laughter, the hushed cries of infants penetrate the darkness and drift across the sand.

Hemiunu is determined that this be the most magnificent tomb on earth: a glorious edifice that will eclipse any tomb ever built—a structure that will cater in splendorous form to Khufu's every need during his passage to the afterlife, that will symbolize the pharaoh's enduring sovereignty on earth, and that will stand as testimony to the power and resplendence of the Egyptian kingdom. In his ambition, he knows that he will see his vision brought to splendid fruition. Yet the people who are building this magnificent tomb will never realize the full measure of their efforts—will never know that history will credit them with erecting one of the most spectacular monuments to human achievement the world will ever behold. Nor will they know that some of their key construction methods will be lost to time, and that the mystery of how they built this great pyramid will puzzle and intrigue humankind for millennia. They do know, however, that this mountainous tomb is rising from the desert perfectly sited, perfectly constructed, and right on schedule as a direct result of their labors, and the knowledge pleases them greatly.

They work in gangs subdivided into teams. The overseers have organized this vast army into dozens of work gangs grouped by trade. The gangs strive to outshine each other in precision and speed, then boast unabashedly when they do. But such boasting is well deserved. Khufu's pyramid is not being built by the uninitiated or the unskilled; it is being built by men who excel at their trades, many of whom worked on the construction of earlier pyramids at Saqqara and Dahshur. Hemiunu would never have attempted construction of this pyramid with anything but a highly skilled and highly trained workforce, and he has assembled the best workers to be had in both Upper and Lower Egypt.

The pace in the quarries has picked up as the need for stone blocks increases, and it is now essential to produce roughly a thousand blocks a day to keep construction on schedule. More than 2 million blocks of stone will be needed to build this pyramid—2.6 million cubic meters of stone weighing 5.7 billion kilograms—and the stonecutters began their

Khufu's pyramid from the knoll.

work early in the construction schedule to stockpile an adequate supply of blocks. The physical demands of this work are incredible. The limestone blocks used in the bulk of the construction weigh between 2 and 6 metric tons apiece, although some reach 16 metric tons; the blocks of granite that will be used to construct the King's Chamber weigh as much as 50 metric tons. Placing the blocks on the pyramid's highest courses is comparable to lifting them to the top floors of a contemporary forty-story building. But these men are unfazed by the challenges of Hemiunu's colossus. They are certain they can lift the blocks higher and higher, and place each and every one of them with the same precision that was applied at the pyramid's base.

How *did* they do it?

Unfortunately, no plans, drawings, or written records regarding the construction of Khufu's pyramid have ever been discovered, and thus the central question of how these men successfully moved all of those blocks into place and stacked them to a height of 147 meters has never been definitively resolved. It is clear from surviving records and from examination of structures built at Giza and before the Giza pyramids that the ancient Egyptians understood the principles of the lever and the inclined plane; that they had a superb grasp of mathematics and spatial concepts and could calculate volumes, slopes, and angles with exceptional accuracy; that they knew how to survey; that they had devised a sound system of measurements based on the cubit; and that they could construct scaffolds. They had no pulleys, which had not been yet invented. Arnold describes "bearing stones," about which a rope could be threaded to make a turn while pulling.[1] It is possible that wooden poles were used as rollers to move objects short distances. Wheeled carts pulled by bullocks appear in tomb paintings from much later periods, but none have been found dating to the Old Kingdom. Thus they had very few available options for moving and placing the stone blocks. They could have constructed ramps based on an inclined plane or planes; they could have jacked up the blocks with levers, fulcrums, and movable supports positioned beneath the blocks; or they could have devised a counterbalance lever system similar to a seesaw.

Given the number and weight of the blocks, the most logical and expedient method was to construct an inclined plane or planes—a ramp or ramps—and drag the blocks upward by means of sheer manual labor. Early records indicate that workers were given bullocks to eat, which confirms that the Egyptians had domesticated cattle, but there was no reference to their use as draft animals, which might have assisted with heavy tasks, until later dynasties. They also had donkeys, which carried loads. But animals would have to be fed and watered—a different type of logistical issue. Therefore I assumed that only human labor was used.

The ramp method had been refined to reduce the degree of physical exertion required and maximize the number of blocks that could be moved per day. For example, workers crafted wooden sledges or rockers to facilitate the handling and maneuvering of the blocks. Available evidence strongly supports the theory that the Great Pyramid's builders devised ramps to move and place the stone blocks. Records survive that reveal the ancient Egyptians knew how to calculate the volume of materials required to construct ramps. But more significantly, I have seen remains of ramps at Giza and other Old Kingdom sites—most notable among them, a large ramp roughly 5.5 meters wide that leads up from the quarry site west of the Sphinx to the vicinity of Khufu's Queens' Pyramids. Clearly ramps were employed by the pyramids' builders in some fashion. (See Plate 16.)

The Rhind mathematical papyrus gives details about calculating the volume of a ramp. Another written record (the papyrus of Anastasi I) includes an exercise asking a scribe named Amenemope to calculate the number of bricks needed to build a ramp 730 cubits (383 meters) long, 60 cubits (31.4 meters) high, and 55 cubits (28.8 meters) wide at the top. This ramp was composed of a series of brick compartments or chambers that would subsequently be filled with earth.[2]

Construction ramps are also shown in tomb drawings. One example, cited by both Clarke and Arnold, is from the tomb of Rekhmira at Thebes. The ramp is constructed of brick, reed mats, and filler material. The top surface appears to be paved with limestone, and on the ramp is a roofing slab in position to be pulled up for placement.[3] Remains of a

Ramps and causeways at Giza and other sites.
TOP: Khafre's pyramid causeway, right background; old quarry, left center;
Menkaure's causeway, left foreground.
BOTTOM LEFT: Ramp to mastaba, west side of Khufu's pyramid, Giza.
BOTTOM CENTER: Ramp retaining walls, south of Khufu's Queens' Pyramids, Giza.
BOTTOM RIGHT: Causeway at the pyramid of Unas, Saqqara.

construction ramp at the east face of a huge pylon at Karnak illustrate one construction technique. Sturdy walls of mud brick were built perpendicular to the face of the pylon, 4 to 6 cubits (2 to 3 meters) apart. Although the walls are in a state of decay, it appears that there were seven of them, giving the ramp a width of 30 to 40 cubits at the top. The space between the walls was filled with hard-packed earth and rubble. As the construction rose, the ramp was extended and the height increased. Ultimately, it would have been about 80 cubits (42 meters) high.[4]

Arnold suggests that construction ramps were built with timber sleepers, or balks, possibly made from old boat lumber. These sleepers were placed on ramps and transport roads perpendicular to the direction of travel, like railroad ties. The spacing was such that the runners of a large wooden sledge would always be resting on at least two or three successive timbers. He includes photographs or drawings of this type of ramp at the pyramids of Amenemhat I and Senwosret I at Lisht, and in the quarry used by Senwosret II at Lahun.[5]

On the east face of the Meidum Pyramid, high up on the fifth and sixth steps, are two distinct vertical lines marking depressions in the finished stonework. They appear to indicate where an inclined ramp leaned against the east face. The ramp height suggested by these marks is 55 meters.[6] Other writers describe the use of ramps at a number of locations.[7] In addition, there are tons of debris on the Giza site, including vast amounts of talfa, the calcareous clay that archaeologists believe may well be fill material from ramps.[8]

In summary, the evidence for the use of ramps—written records, ramp drawings, and the physical remains of dozens of ramps built in several different styles—is overwhelming and compelling.

But what *about* the alternatives? Could the builders of Khufu's pyramid have moved and placed the stone blocks by other means? Not unless they had developed some type of highly advanced technology or equipment that vanished into history without a trace. Khufu's pyramid is not only enormous, it is also intricate in plan. Its upward progression involved much more than the stacking of block upon immense block in a perfect pyramid formation; it entailed a complex integration of interior and exterior architectural schemes, parts of which involved the placement of massive beams and blocks of stone weighing upwards of 50 metric tons. It is hard to imagine that any of the alternative methods could have provided this integration. And even if these methods had somehow been combined with some other method that enabled workmen to construct the interior of the pyramid, the use of a jacking system or counterbalanced lever system would have been impractical for a construction effort of this magnitude. Imagine trying to balance a 3-ton block of stone on blocks of wood, then levering one side of it up another 2 or 3 centimeters. Now suppose you are doing this thirty stories in the air and the block slips: carnage would ensue. The seesaw concept is intriguing, but it would have involved a huge quantity of good lumber, which was in short supply in ancient Egypt. No illustrations of this concept have ever been found, and there are no archaeological remains that indicate lumber was used on a large scale in this construction.

Concluding that the Great Pyramid's builders moved and placed the stone blocks by means of a ramp or ramps does not solve the mystery of how this pyramid was built, but it does provide the key with which to unlock the mystery: the fundamental means devised by the ancient Egyptians to cope with the extraordinary height and mass of the structure. No concrete evidence remains, however, to tell us whether they constructed a single ramp or many, or to tell us—if they did in fact use multiple ramps—how many there were and how they were configured. For the answers to these questions, we must first consider why a ramp or ramps were needed.

Khufu's pyramid posed a number of formidable challenges to its builders, but the ultimate challenge in erecting this structure was its astonishing height. The Bent Pyramid and the Red Pyramid, each 200 cubits high, were the largest structures built up to that time. (See Plates 6 and 7.) But they pale in comparison to Khufu's tomb, which at 280 cubits is 80 cubits higher than these structures. Without the last-minute change of slope, the Bent Pyramid would have been 250 cubits high. Given this historical progression in heights, I believe the challenge posed by Hemiunu was to exceed 250 cubits in height.

I consider it unlikely that Khufu would have been content to build his tomb on the same scale as his father's, but the extent to which he may have dictated the height of the pyramid may never be determined. The point is that the height specified for this structure was quite simply fantastic. Hemiunu had clearly absorbed the lessons of Dahshur. He no doubt summoned the architects, foremen, and crew chiefs who had erected Sneferu's tombs to discuss their experiences, entertain their suggestions, pick their brains. But there was more to the creation of Khufu's Great Pyramid than improving upon the pyramids built for other pharaohs. Hemiunu's objective was to create a pyramid for Khufu that would rise from the sands of Giza as a structure without equal, one that would spire into the clouds as a perfect glorious symbol of the relationship of the pyramid and the pharaoh to the sun god.

In the end, Hemiunu achieved his goal. With a height of 280 cubits, Khufu's pyramid surpassed anything previously built. Even the pyramid built next door by Khufu's son Khafre falls short of this height. And for the next 4,000 years, no other structure attained this stature. Hemiunu accomplished his objective by designing a foolproof means of moving ton upon ton of stone in an upward progression that effected flawless architectural and structural integrity.

Having established the height of the pyramid, the next decision the ancient designers had to make was determining the slope of the sides. This decision would influence not only the quantities of building materials required to achieve the desired height, but also the construction methods. There is a clear trade-off here. The steeper the slope of the pyramid faces, the easier it is to achieve great heights with less material, but also, the greater the structural design issues and the more difficult the construction challenges. Think of it as building a highway in alpine mountains. The risk of accidents and the sheer effort to lift materials increase as the slope of the work surface becomes steeper.

Khufu's pyramid is architecturally complex, and the construction of the interior elements added many challenges because of the size and positioning of the massive blocks and beams used. The interior incorporates the Descending Corridor, the Lower Chamber, the Ascending Corridor, the Grand Gallery, the Queen's and King's Chambers, several interconnecting tunnels and corridors, and a network of shafts high in the body of the masonry and oriented to the northern polar stars to ensure the smooth passage of Khufu's spirit to the heavens.

The builders of earlier pyramids lacked the confidence to successfully construct interior chambers that would not collapse under the massive load of stone overhead, so the burial chambers were carved into the bedrock beneath the pyramid. Not until Sneferu built the Bent Pyramid and the Red Pyramid at Dahshur were the first tentative efforts made at placing a chamber within the structure. Under Hemiunu's oversight, the designers perfected the technology of interior structures in Khufu's pyramid to a degree never duplicated. Thereafter, burial chambers were again constructed on or in bedrock beneath the pyramid.

Given the height, slope, and complex interior configuration of Khufu's pyramid, ensuring its structural integrity entailed some system of incremental reinforcement and some infallible means of elevating men and materials from level to level—a ramp or system of ramps. But what did this ramp or ramps look like? What about its height, length, width, angle of inclination, position? The best way to answer these questions is to re-create the construction sequence, the various junctures at which a ramp or ramps would have been needed, and how they would have facilitated the construction process.

These construction details would have influenced the ramp design. The height of the course was determined by the size of the casing stones. Obviously, the larger the stones, the fewer required. But larger stones also involve more work and are more difficult to place, particularly at the higher courses. The size and required delivery rate of stones would influence the width of the ramp or ramps needed to provide adequate space for maneuvering. The more blocks that must be delivered per day, the wider the ramp must be to allow multiple teams to ascend the structure. Therefore the next step in designing the ramps was establishing how many blocks of stone would be required to construct the pyramid.

From the known dimensions, the total apparent volume of the pyramid (including the original top courses, which are no longer present) can be found as follows: volume = 1/3 x area of base x height. This yields 0.333 x 230.4 x 230.4 x 146.6 = 2.59 million cubic meters.

Petrie's survey (conducted in the 1800s) included measurements of the height of each course, taken at several different points, as discussed in Chapter 3. There has been considerable weathering of the limestone in some areas, so I believe the older data are probably more accurate than any that could be gathered today. When graphed, these measurements show an intriguing profile of block heights up to Course 203 (see graph on page 103). Although the top 10 meters or so are missing, there were actually about 218 courses, including the pyramidion or capstone.

The width of the exterior blocks can be measured, but not the length—except in a very few instances. A Course 1 block that I measured on the east face had dimensions of 149 centimeters high, 212 centimeters wide, and 230 centimeters long. Marks in the quarry also provide a clue to the height, width, and length of the blocks. To estimate the number of blocks I used two sets of ratios. For blocks more than 1 meter thick, I used a height:width:length ratio of 1.0:1.33:1.5, while for the balance of the stones I assumed a ratio of 1.0:1.33:2.0.

The mass of bedrock that rises up from the plateau and was incorporated into the structure can be seen at the northeast corner, where it rises 6 or 7 meters. The precise dimensions are unknown.[9] The volume of blocks required is therefore reduced by a corresponding amount. In my analysis, I assumed the bedrock protrusion is a rectangle 150 meters long by 150 meters wide and 10 meters high, or 225,000 cubic meters.

To determine the weight of the blocks, I analyzed limestone samples at Giza and found density in the range of 1.8 to 2.2 grams per cubic centimeter. Engineering books report 1.8 to 2.7 grams per cubic centimeter for limestone (the higher values are for dense limestone). A value of 2.2 grams per cubic centimeter (2,200 kilograms per cubic meter) has been reported by several respected researchers at Giza, so this was the value I used.[10] To determine the approximate number of casing stones, I assumed that they were the same width and height as the blocks on each corresponding course and were cut at an angle of 51.9°, the angle of inclination of Khufu's pyramid.

With this information, it is an exercise in trigonometry to construct a computer model that will calculate the length of the sides and therefore the area of each course or level of the pyramid, once the height of the block is known. Starting from the base dimensions of 440 by 440 cubits, the program calculates the length of a side at the next course. Then, at a new height equal to the height of that course, it repeats the calculation and finds the new length. In other words, the model makes what is called a "rise and run" calculation. The model is formulated to calculate the number of blocks per course as well. For the lower levels, I deduct the volume of the interior chambers and corridors and make an allowance of 225,000 cubic meters for the bedrock protrusion. The model shows that the pyramid contains 2.2 million blocks of stone in the core and that 98,000 casing stones are required.

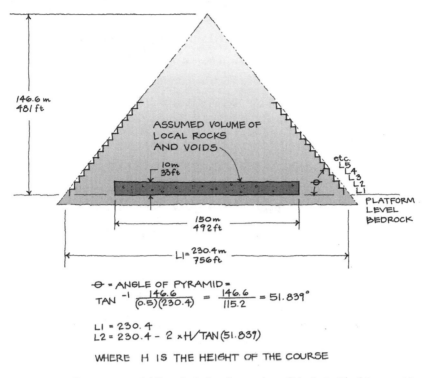

$$\theta = \text{ANGLE OF PYRAMID} =$$
$$\text{TAN}^{-1} \frac{146.6}{(0.5)(230.4)} = \frac{146.6}{115.2} = 51.839°$$

$$L1 = 230.4$$
$$L2 = 230.4 - 2 \times H/\text{TAN}(51.839)$$

WHERE H IS THE HEIGHT OF THE COURSE

Computer model for calculating the number of blocks in Khufu's pyramid.

By having the model calculate the weight of the block and the height of the course where it is placed, the labor to place the block can be estimated later. These calculations are illustrated in Appendix 3. They reveal many interesting aspects of the pyramid construction, some of which are excerpted below.

Excerpts from Rise and Run Calculations

COURSE No.	HEIGHT (M)	No. OF BLOCKS	BLOCK WEIGHT (KG)	CUMULATIVE No. OF BLOCKS	PERCENT OF TOTAL
I	1.5	6,313	14,656	6,313	--
II	11.7	23,598	3,493	120,901	5
25	22.5	20,984	3,247	507,355	23
55	46.4	22,693	1,535	1,096,714	50
76	61.0	17,978	1,359	1,453,902	67
100	77.7	5,472	4,094	1,767,121	81
218	146.6	1	1,973	2,186,053	100

By the time Course 55 is reached, one-half of the blocks are installed. This is only 46 meters of the 146-meter height of the pyramid, or 31 percent of the way up. By Course 76 (61 meters, or 42 percent of the way up), two-thirds of the blocks have been installed. At the higher levels, fewer stones are required, and they are smaller. This finding has important implications because it suggests that ramps can be narrower—and possibly steeper—on the upper levels of the pyramid.

To finalize the actual number of blocks, we need to adjust for the interior void spaces. These are estimated from published dimensions (converted to meters).[11] Including the various corridors, chambers, and the Grand Gallery, I found that the volume of the void spaces was 2,096 cubic meters.

Above the Queen's Chamber, the King's Chamber, and the Grand Gallery, special structural beams further subtract from the number of blocks needed. If we assume they are around 1 meter thick and overlap the chamber walls 1 meter, they would add another 333 cubic meters. Add another 600 cubic meters as an allowance for anything else, and the total equivalent void space is around 3,000 cubic meters.

The volume occupied by 2.2 million blocks is 2,590,000 cubic meters less the bedrock protrusion of 225,000 cubic meters, giving a net volume of 2,365,000 cubic meters, or 1.08 cubic meters per block average. Thus the chambers, tunnels, and other elements would eliminate around 2,800 blocks from the calculation described above. This is insignificant in view of the assumptions made in the analysis. For our purposes, then, we can assume that the construction challenge is to place 2.2 million core blocks while constructing the corridors and chambers in the interior of the pyramid, and then placing the 98,000 casing stones.

Various ramp theories are known; they assume multiple ramps to each face, a long ramp to the top, a square helical ramp encircling the pyramid, etc. Known ramps had slopes of 8° to 12° (grades of 14–21 percent), although some roads of 14° to 20° have also been reported. Edwards reports slopes of 1:8 to 1:12.[12]

Now we can explore the question of what the ramps might have looked like. There could have been a single large ramp, different ramps to different faces, ramps enveloping the pyramid, or ramps supported by the pyramid structure itself, rather than being supported by the ground. The first consideration is the slope of the ramp. The remains of ancient ramps can be seen at various pyramid sites in Egypt with slopes of 8 to 12 degrees. When discussing the slope of roadways, highway engineers use the term *grade*: the tangent of the slope angle expressed as a percent. Conservative highway design attempts to limit grades to 6 percent or less (it can be 9 or 10 percent if the distance is short), which means that the road rises or falls 1 meter for every 16.7 meters of forward travel. For the design of the main ramp, I assume the maximum slope would be about one unit of rise for every six units of forward travel (an angle of about 9.5°). This corresponds to a 16.7 percent grade in highway design. Anything steeper than this would be difficult to use and hazardous if loads broke free.

Various ramp concepts

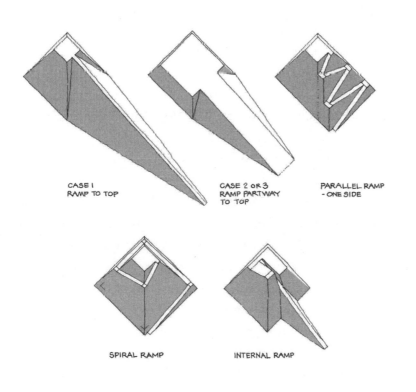

CASE I
RAMP TO TOP

CASE 2 OR 3
RAMP PARTWAY
TO TOP

PARALLEL RAMP
- ONE SIDE

SPIRAL RAMP

INTERNAL RAMP

The second consideration is very simple, and since the ancient Egyptians were practical people, I believe they would have endorsed it: construction of the ramp must involve less material, less volume, and less work than construction of the pyramid itself. With that in mind, one can quickly demonstrate that a single long ramp to the top of the pyramid would not be the solution (Case 1). Such a ramp would be more than 800 meters long (more than half a mile) and require in excess of 8 million cubic meters of material—more than three times that required for the pyramid itself!

Since half of the blocks have been put in place by Course 55, a large ramp to the height of Course 55 would be more practical (Case 2). A ramp of this size would be 46.4 meters high and 278 meters long with a 16.7 percent grade, the assumed upper limit. This ramp would have a volume of 750,000 cubic meters, or 29 percent of the volume of the pyramid itself. To reach Course 76, where two-thirds of the blocks have been installed, the ramp would have to be 61 meters high and 366 meters long, or equivalent to 45 percent of the volume of the pyramid (Case 3).

At this point, we have considered three configurations for ramps. Case 1 (to the top) is ruled out by the sheer volume of work involved. Cases 2 and 3 require auxiliary ramps to reach the upper levels of the structure. Case 2 needs five auxiliary ramps to reach Course 194, about 12 meters below the top of the pyramid. To reach this same point, Case 3 needs four auxiliary ramps. In Case 3, the added work of constructing a higher main ramp is partially offset by the labor saved by eliminating one of the auxiliary ramps.

The ramp dimensions are also influenced by the construction schedule. For the pyramid builders to adhere to a reasonable schedule, the ramp must have been wide enough to allow multiple teams to approach the working surface, deliver their loads, and leave without hampering the work of other teams. Case 2 yields a ramp width of 74 meters at Course 55. This width enables twenty-four stone-hauling teams to approach Course 55 simultaneously, given the size of the stones they would be hauling. At Course 9, thirty-four teams could approach the working level simultaneously. Here the width of the ramp is 136.5 meters, but the larg-

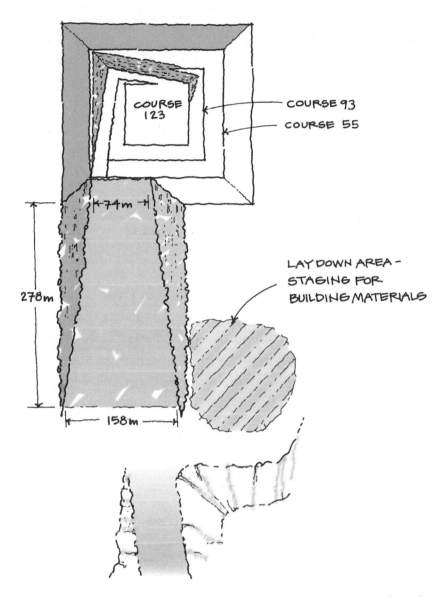

COURSE 123

COURSE 93
COURSE 55

←74m→

278m

158m

LAY DOWN AREA –
STAGING FOR
BUILDING MATERIALS

Primary ramp (Case 2).

er blocks require larger hauling teams, so fewer teams can be accommodated at one time.

Another way to examine the placement of ramps is to consider the
level at which the construction of the interior features takes place. The
bottom of the King's Chamber is at an elevation of 43 meters (Course
50), while the top of the King's Chamber is 49 meters above grade

(Course 59; see drawing on page 92). At these elevations, the top of the pyramid would have been a flat, square surface larger than a football field. This surface would provide a large working area for special-purpose auxiliary ramps needed to raise the heavy blocks and slabs used to frame the Grand Gallery, the King's Chamber, and the stress-relieving chambers above the King's Chamber (see illustrations on pages 98 and 99). Based on these considerations, a single ramp to Course 55 would provide an excellent working platform for construction of the King's Chamber. The five Relieving Chambers above the King's Chamber are 21 meters above the floor of the King's Chamber, which would put the top roof beams at an elevation of 63 meters (Course 80). This work could have all been put in place while working from the level base provided by Course 55, its broad ramp providing easy access. A temporary ramp wide enough to maneuver the 7-meter-long granite beams used in the ceiling of the King's Chamber and the beams in the Relieving Chambers could then be built next to the King's Chamber. As the beams in the Relieving Chambers were placed, additional stonework in the form of bulwarks of masonry erected on top of Course 55 would provide

Auxiliary ramp concept.

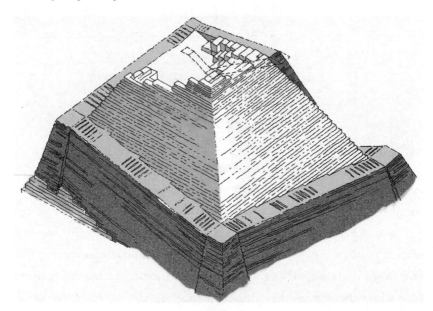

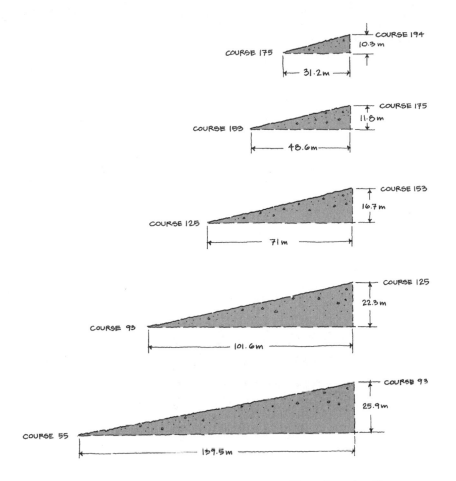

COURSE 194
10.3 m

COURSE 175

31.2m

COURSE 175
11.8 m

COURSE 153

48.6m

COURSE 153
16.7 m

COURSE 125

71m

COURSE 125
22.3 m

COURSE 93

101.6m

COURSE 93
25.9m

COURSE 55

139.5m

Dimensions of auxiliary ramps.

lateral bracing for the chambers. Once the structure of the Relieving Chambers was complete, this temporary ramp could simply be incorporated into the body of the pyramid as it was raised and filled in around the chambers.

To continue construction and reach the level above the Relieving Chambers from the working platform at Course 55 (46.4 meters high), an additional ramp 7 meters wide would be needed on top of the existing construction (along the west edge). It would have the same slope and extend up to 72.3 meters (Course 93). This ramp would then turn the corner, progress up the north side of the pyramid, and be extended to Course 125, where the height is 94.6 meters. The diagram above illustrates the approximate dimensions of these auxiliary ramps. Note that since fewer blocks are required, the delivery rates are reduced, and the ramp width can be decreased gradually from 7 meters to 2 to 4 meters in width on the highest levels.

From this point on Course 125, three more "wraps" of the pyramid would be made from Course 126 to 153 (east side); Course 154 to 175 (south side); and Course 176 to 194 (west side). These ramps are smaller and steeper than the main ramp. At this point, the elevation is 133.4 meters, with about 12 meters remaining to the top platform, where the final pyramidal capstone would be put in place.

The concept of a spiral ramp wrapping around the pyramid and supported by the stair step structure of the pyramid itself seems to make the most sense. It is the most economical approach in terms of building materials and lends itself to gradual extension as each succeeding course of masonry is placed. Also, at higher levels the size of the stones decreases significantly. The builders obviously had no problem cutting and handling larger stones, since the lower levels of the pyramid are constructed in this manner. It is also obvious they were not limited by materials,

Construction of auxiliary ramps.

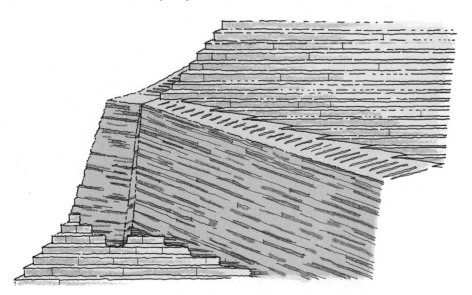

because even today one can find limestone layers in the old quarries at Giza where large stones could be cut. Larger stones have the advantage of greater stability, and of course the volume is completed more quickly, which requires less masonry work.

So why would smaller stones be used up high? I believe the explanation is that smaller stones were simply easier to handle near the top. They permitted the use of narrower, steeper ramps. Fewer men were required to put them in place. Smaller stones could effectively have been manhandled into position near the top, something not possible with larger stones, which would have afforded the masons greater flexibility.

Still unanswered is the question of that final 12 meters. How did the builders place the topmost stones? Here additional ramps seem less plausible, because the pyramid is so narrow at the top that little gain is made in elevation at reasonable slopes. At Course 194, the pyramid is a scant 23 meters wide. Also, at this level the height of the blocks is around 0.5 meters, and their weight is 800 to 1,000 kilograms, suggesting that the final 3,700 blocks could be stacked up by workers constructing a staircase in the structure and using it to manhandle the blocks into place.

Case 3 (a ramp to Course 76) provides an alternate scheme for transporting the heavy beams for the roof of the King's Chamber and the Relieving Chambers up to the required elevations. I considered a modified Case 3 in which the width of the ramp was decreased to 10 meters at the top. This variant reduces the volume of the ramp to 27 percent of the volume of the pyramid itself but reduces the ramp width at Course 55 to 50 percent of that in Case 2, and to 70 percent of the width of that in Case 2 at the lower levels (Courses 25 and 9).

Reverting to the Case 2 ramp, another advantage of the broader ramp to Course 55 is that the top of the ramp can be used as a staging area for stone delivery to the higher courses. At the higher levels, the size and quantity of the blocks decreases, and smaller teams can deliver them. A supply of blocks can be brought up and stockpiled at the upper end of the ramp. From this point, delivery crews can move them up the auxiliary ramps, which shortens the travel time for the crews working on the higher courses.

No doubt other alternatives or variations in ramp schemes can be considered. But there is still one other variable to be evaluated—the capability to deliver stones at a sufficiently rapid pace to complete the construction in a king's lifetime. To assess this variable, one has to choose a ramp configuration.

Ultimately, I settled on a hybrid scheme that is a combination of several approaches. From a main ramp that extends up to Course 55 (Case 2), I envision a series of auxiliary ramps wrapping around the pyramid, up to Course 194. The difference is that these ramps are much narrower and are supported by the pyramid itself rather than built from the ground up. Therefore, they require less material to construct. At the corners, additional blocks stacked on the existing lower courses form a platform to create a wide enough takeoff point for the auxiliary ramp, and likewise at the next corner to make room for the sledge to turn the corner. These ramps extend to an elevation at which the horizontal distance is long enough for a significant gain in height. Above this point, a different method is needed.

At the very top, a "staircase" is left in the center of the construction so the blocks can be pushed from below and pulled up by ropes over poles or bearing stones—and then put in place. This staircase makes it possible to position the capstone. Once the capstone is maneuvered into place, the staircase is filled in. An alternate approach would be to bring the capstone up to the last level reachable by a ramp, then jack it up as the balance of the pyramid is constructed: as the pyramid is built beneath it, it rises with the remaining levels.

The construction of the intricate interior structures seems less onerous if one concludes that a large ramp extends up to a flat working surface, where these interior features are built. The ascending tunnel and upper chambers are built as the rising pyramid is constructed around them. Once the level of a chamber floor is reached, the structure (walls and roof) is placed by working from the flat platform of the pyramid's top surface. Then the walls are erected, the pyramid structure is placed around them, the chamber is roofed over, and the structure continues to rise—a pyramid within a pyramid, so to speak.

"Staircase ramp" construction.

As mentioned, the delivery rate for the stones is a final consideration in determining the size of the ramps. Roughly speaking, it is necessary to deliver 1,000 stones per day just to build the pyramid (excluding the interior structures and other facilities associated with the pyramid complex.) A better estimate can be made by determining the time required to perform each individual task, and from this the size of the workforce (Chapter 8). At this point we just need to get a feel for what stone delivery rates can be practically achieved with the ramp scheme postulated above to determine if it is indeed workable.

The main ramp has a height of 46.4 meters and a width at the base of 158 meters. The sides are sloped at an angle of 48° to provide adequate structural capability without excessive use of materials. This ramp requires a little less than 750,000 cubic meters of material (about 29 percent of the volume of the pyramid itself). The size of the ramp is dictated by both practical engineering considerations and the size of the workforce. The construction schedule establishes the rate of production of stone blocks from the quarry, which dictates the number of laborers needed to haul stones up the ramp, which in turn determines the size of the ramp needed. However, in this as in all similar construction operations, there are trade-offs. First, consider the size of the teams needed to pull the blocks up the ramp, which is a function of the block size. I assumed the heaviest blocks were moved at a speed of 2 meters per minute, those weighing less than 3,000 kilograms at 3 meters per minute, and the smallest blocks at 4 meters per minute:

Workers Needed to Haul Blocks Up the Ramp

BLOCK SIZE (kg)	TEAM SIZE (workers)
14,000	84
8,000	48
6,000	36
4,000	24
2,000	12
1,000	6

The ramp width and length will determine how many teams can be working on the ramp at one time. The largest teams are required only for the first three courses. Thereafter the team size required to haul a block is reduced, until it is time to place the heavy granite beams that form the superstructure of the King's Chamber. As a rule, more stones are placed at the lower levels, and they are larger and heavier, while fewer, lighter stones are placed at the higher levels. This does not apply at every level, however. The stones are so large on the bottom three courses that fewer are required, and the bedrock protrusion further reduces that number. At certain higher levels, fewer and heavier stones are used to reinforce the structure. Up high, the number of stones increases because they are smaller, and smaller teams are needed to haul them. The configuration of a typical hauling team is shown in the drawing.

Typical hauling team and sledge arrangement in plan view.

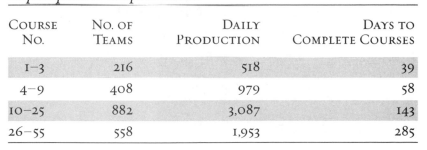

A sledge similar to this one was found at Dahshur. It is 4.2 meters long and 0.8 meters wide, and can be seen in the Cairo Museum. Sturdily assembled with wooden pegs and dovetailed joints, it is a good example of the Egyptian carpenters' craft. It has various notches and holes for attaching and securing loads.[13]

A "practical" ramp—one that compromises between delivery capacity and construction time and effort—would not be large enough to deliver all of the large stones needed to maintain a constant production rate, although it would be more than adequate to deliver smaller stones at the higher levels. The following table shows the number of teams, the daily production of stones, and the days required to complete that course at the given production rate. "Daily production" means stones delivered to the indicated course by the hauling teams. I assume that the stonemasons working on each course could place the stones as fast as they could be delivered:

Days Required to Complete Courses

Course No.	No. of Teams	Daily Production	Days to Complete Courses
1–3	216	518	39
4–9	408	979	58
10–25	882	3,087	143
26–55	558	1,953	285

The ramp is constructed by first laying out the full width of 158 meters against the base of the pyramid after the first course has been placed. Retaining walls of stone and mud bricks are constructed, and the roadway is filled in with rubble, clay, sand, and chips of limestone. At first the ramp resembles a broad roadway. Laborers use it to bring backing stones up onto the platform. Starting at one of the corners (say, the northeast corner), they are put in position and carefully aligned, first along the north face, and then along the east face. Meanwhile, other backing stones are moved up the ramp and placed on wooden rollers at various locations on the platform, ready to be moved into final position. Additional rollers are stockpiled near them.

At the appropriate time, laborers move these backing stones into position, and the stonemasons dress them down to a finished dimension for an exact fit to the adjacent block. Mortar is applied as a lubricant to the base surface before the stone is pried into its final position. Once the entire northern face is lined with backing stones, and once they are also in position partway along the east and west edges, then the area behind them can be filled in with interior stones. Here the work need not be so precise. Odd sizes and shapes are likely to be used; small crevices and gaps can be filled with chips, sand, or mortar if required.

As each course nears completion, the ramp crews begin stockpiling long rectangular stones to raise the ramp retaining walls in preparation for elevating the ramp to the next level: first one side, then the other.

This method permits the hauling crews to switch over and begin taking stones up to the next level while the ramp crews finish raising the other side of the

Wall at Saqqara, showing mud-brick construction.

ramp to the new level. Ramp construction proceeds hand in hand with pyramid construction. The structure rises from the desert plateau bedrock like some gigantic beast, changing form daily. Meanwhile, gangs of workers pass in a continuous stream, dragging blocks up the ramp on wooden sledges. They have created a series of tracks in the ramp that the sleds ride on, and they keep these tracks lubricated so that the sleds move freely but the laborers have good traction where they walk. If only we could go back in time and observe this extraordinary spectacle!

Under Hemiunu's guidance, and with the able crew he has assembled, the pyramid continues to rise, soaring to the sky. After all of the work is completed on the King's Chamber and the Relieving Chambers, work begins on the first auxiliary ramp.

From this point on, construction proceeds using auxiliary ramps supported on the pyramid structure itself. The first auxiliary ramp extends from the working platform at Course 55 along the west side of the pyramid up to Course 93, a rise of 25.9 meters. A small section of blocks is added to Courses 50 to 55 at the corner to provide a foundation for this ramp, which is constrained by a retaining wall initially built along the edge of Course 50. As each new course is added, the ramp is extended up to the next level. After several courses it reaches its full width of 7 meters without need for extra blocks. Once the structure has reached Course 93, the ramp turns the corner, and the process is repeated on the north face, the auxiliary ramp now extending from Course 94 up to Course 125. These ramps are lined with timber sleepers spaced 2 to 3 cubits apart to minimize friction.

At these heights, safety is critical. We have no formal record of accidents, but an informal record can be established from the bones unearthed in the workers' tombs by Hawass. Numerous fractures were skillfully set. What will never be known is how many workers died during construction. One can imagine how a group of laborers might be distracted by some event in the valley or, in a careless moment, lose control of their load. Before they recover, a stone plunges over the edge along with its sledge and several stone haulers, who are seriously injured or killed by the fall. As the construction proceeds higher, the risk of acci-

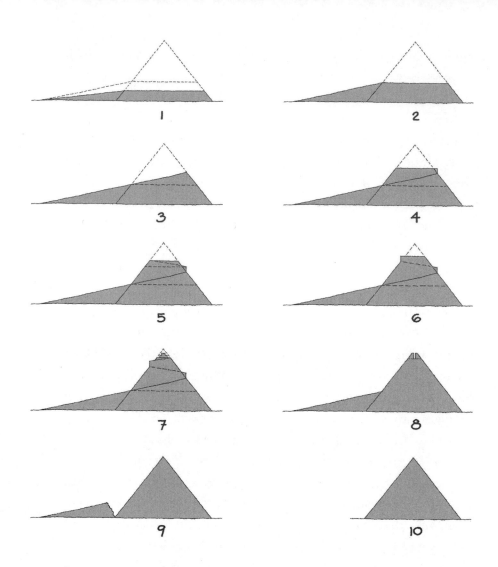

Ramp construction sequence.

dents increases, and more skill is required to ensure that the slope and measurements remain true to Hemiunu's scheme.

Above Course 125, three more wraps of the pyramid bring the height to Course 194, where the elevation is 133.4 meters. At this level the ramp width is 2 meters, and the distance to the top platform, where the pyramidion is put in place, only 12 meters. This final step uses the staircase approach described earlier. Once the pyramidion is in place, the casing stones are installed around the topmost layers.

Vertical face of quarry with old ramp debris.

The pyramid is finished with white limestone casing stones that are triangular in cross section and fitted into the "stair step" outer faces of the backing stones to provide a uniform, smooth exterior surface. At the topmost levels, scaffolding is erected to place these blocks, and the work proceeds downward course by course. Because these blocks are half the size of the rectangular blocks (around 210 kilograms), they can be wrestled into position by a crew of six. Once they are placed, masons trim them to the final dimensions to create a smooth surface with the correct angle of inclination. Also at this time, the missing blocks are filled in as the "staircase" is removed, and any final trimming is performed. As the casing stones are placed and finished on the lower levels, the ramps are gradually demolished and the debris hauled away to fill some of the voids created by the quarry operation.

Today, if you walk to the edge of the former quarry area, you see embankments 10 meters high of loose soil and sand, scraps of broken pottery, chips of limestone, and other debris. Almost certainly the fill material used in the ramps, it was disposed of by carrying it to the closest point and dumping it over the edge of the former quarry.

Only one other structure built in the Old Kingdom can compete with Khufu's pyramid in terms of size, grandeur, and engineering complexity—the system of ramps that was erected to build the pyramid itself. Regrettably, this remarkable example of Fourth Dynasty ingenuity and skill was obliterated as the final measure before Khufu's stair step to the gods was consecrated.

CHAPTER 7

STAIR STEPS
TO THE GODS

Let us now return to the Great Pyramid's construction sequence. When we left off, the site was graded and leveled and the first courses of masonry were being placed.

Prior to the construction of the Red (North) Pyramid, the burial chamber was always constructed in or on bedrock beneath the pyramid. Perhaps it was the strength of this long-standing tradition that led Hemiunu to order the construction of the Descending Corridor and the Lower Chamber. Or the Lower Chamber may have been important for religious reasons. Its true purpose is not known. In any case, the chamber was partially constructed, a new tunnel was started, presumably leading to a second chamber, and then the work was abandoned. This was a prodigious effort to be undertaken and then discontinued.

Consider the Descending Corridor itself. Workers opened up a passageway barely 1 meter wide and 1.2 meters high. Only a child or a midget could stand in that opening; an adult would be forced to work in a stooped position. The corridor descends at an angle of 26.5°, becomes horizontal, and then extends for an additional distance before opening up into the Lower Chamber.[1] The total volume of rock removed was in excess of 500 cubic meters. It is difficult to imagine that more than two persons could fit in the opening and have access to the stone face at the same time. While they chiseled away at the limestone, one or two other laborers behind them would be picking up pieces of stone and chips and placing them in baskets for removal by a human chain of laborers. Other laborers on the surface would carry away the baskets to keep the work site cleared of rubble. It would take several minutes to pass each basket load (0.025 cubic meters) up to the surface, dump it, and return it to the masons working below. I estimate that two masons could cut approximately 1 cubic meter of limestone in a day in the quarry. Underground, with little light, poor ventilation, and cramped working conditions, the production rate was probably slower, and it may have taken 100 days to excavate the corridor. Once the chamber was opened up, additional masons could work on it, so the pace there could have been faster, suggesting that the entire project was completed in a year to a year and a half.

A word about the angle of declination: There have been attempts to correlate this angle with certain stars that might have had some significance in the Egyptian religion. I believe there is a far simpler explanation. First, the ancient Egyptians did not measure angles or slopes in degrees but instead measured slope as horizontal setback per rise of 1 cubit—the seqed. An obvious and convenient measure would be a slope of 2:1, which makes field calculations by the masons simple to perform and implement. In modern terms, the angle equivalent to a slope of 2:1 (or a tangent of 1/2) is 26°34′, and the corridors in the pyramid are built as closely to this measurement as the masons were able to achieve, given the limitations of their tools and the inevitable problems of building with normal construction tolerances. Therefore I believe that all of the corridors except the horizontal ones were intended to have a slope of 2:1.

Above ground, the Descending Corridor is neatly framed with limestone that had to be put in place as each successive course of the pyramid was constructed. The Descending Corridor rises to Course 19 (17 meters above grade), where the original entrance to the pyramid is located. This opening is framed with massive beams of stone, including four placed in a V-shaped arch to serve as a lintel over the opening. At Course 19, an abrupt increase in the height of the course accommodates the opening. In addition, the next two courses (20 and 21) have a small height so the masonry can better be fitted around these beams. Following Khufu's burial, this opening was sealed and covered with casing stones, so all traces of it were concealed.

Approximately 28 meters below the entrance, another corridor, rising at 26.2°, extends upward 39 meters to the point at which the Grand Gallery commences. The Ascending Corridor has the same dimensions as the Descending Corridor: 1.05 meters wide and 1.2 meters high. One difference is that the Ascending Corridor is framed with girdle stones, rectangular blocks with an aperture equal to the size of the corridor cut through them. The girdle stones are uniformly spaced up the corridor every 10 cubits (5.24 meters) above the point at which the granite blocking plugs were to be placed. The girdle stones were obviously intended to tie the Ascending Corridor into the layers of masonry and resist the downward thrust that would otherwise result. The Descending Corridor had no such need, because the downward thrust of its masonry is resisted by bedrock.

The Ascending Corridor rises to a point 21 meters above grade (Course 23). Where it joins the Grand Gallery, the Horizontal Corridor, 33 meters long, leads to a small antechamber and then into the Queen's Chamber.

Except for the construction of the Ascending Corridor and the Descending Corridor in the interior of the pyramid, the work on Courses 3 to 22 consisted of placing nearly 500,000 stone blocks. In the first three courses, these blocks were large—from 1.2 to 1.5 meters high—and weighed from 10 to 15 metric tons. In the next four courses, the weight of the blocks dropped to 6 to 8 metric tons, and in the next five courses, to 4 metric tons (except for Course 9, which reverts to 7 metric tons). In Courses 13 to 21, the blocks weigh 1.5 to 2 metric tons, except for Course 19, where they revert to 5 tons (see Appendix 3 for block sizes and weights). This pattern is repeated in a similar (but not identical) manner throughout the structure. Periodically, bands of larger blocks are placed to strengthen the overall structure.

I believe these bands of larger stones are significant structurally and in terms of constructability. They are repeated at more or less regular intervals and appear to be associated with the major internal features. For example, the band at Course 19 is associated with the entrance to the pyramid; at Course 22, with the floor of the Queen's Chamber; at Course 44, with the approximate midpoint of the Grand Gallery; at Course 59, with the roof of the King's Chamber; and at Course 74, with the top of the Relieving Chambers.

Fifteen major bands are spaced from 10 to 20 cubits apart (averaging about 15 cubits apart in spacing). In addition, three intermediate bands (Courses 22, 47, and 59) are respectively 4, 5, and 6 cubits above the bands at Courses 19, 44, and 47 and appear to be spaced more closely for strengthening. The courses corresponding to these bands appear to be in uniform numbers of cubits above the platform base: for example, the floor of the Queen's Chamber, 41 cubits; the floor of the King's Chamber, 82 cubits; the top of the Relieving Chambers, 122 cubits. Above the Relieving Chambers, the bands appear to be placed at regular intervals:

Course Elevations in Cubits

COURSE NO.	NOMINAL ELEVATION (cubits)	CALCULATED ELEVATION (cubits)
118	170	171.2
130	186	186.2
144	202	202.6
164	224	224.0
180	240	240.2
196	256	256.8

These conclusions are based on Petrie's measurements from the northeast and southwest corners. His measurements showed that the blocks were rarely the same height, differing by as much as ± 5 inches (0.24 cubits). The variations were much greater in the lower courses (say below Course 100) than in the upper courses, so I averaged his two measurements to determine the height of each course, as shown in Appendix 3. Petrie's block height measurements deviated from the average value by as much as ± 14 percent, and the average variation was about ± 2.67 percent. These values correspond to variations in the height of a course ranging from ± 0.04 cubits to ± 0.21 cubits. While some of these differences tend to cancel out, over fifty courses or more the cumulative difference could amount to several cubits. Also, while it is logical (and consistent with observed practices) to believe that the ancient Egyptians placed the important features of the pyramid at elevations measured in whole cubits, expediency may have led to the use of blocks that were not cut exactly to size. Normal construction tolerances would lead to some variation in measurements, but there are blocks with heights of 2.5 cubits (Course 2), 2.0 cubits (Courses 7, 36, and 44), 1.5 cubits (Courses 22, 30, 40, 41, 68, 69, 118); 1.25 cubits (Courses 32, 33, 34, 51–54, 61, 63–65, etc.), and 1.0 cubit (Courses 155–95). Finally, with the passage of 4,500 years, differential settlement and earthquakes inevitably have affected some of the pyramid's dimensions, so an exact determination of the as-built dimensions is impossible.

Where the dimension is close to an even number of cubits, within the error ranges described above, the original design probably intended for the dimension to be an even number of cubits. Nowhere is this more evident and convincing than in the floor plans of the Queen's and King's Chambers. For this reason, in this chapter I give many of the principal dimensions in cubits (see Appendix 2 for conversion factors).

Consider Course 9 as an example of how large stones were handled in the lower courses. Here the ramp is 136 meters wide, and the hauling distance from the staging area to the center of the pyramid is 413 meters. A ramp this size can accommodate up to 34 teams side by side, each team consisting of 42 laborers, and 12 teams in sequence, meaning that 408 teams could theoretically be hauling stones at any given time. At a travel speed of 2 meters per minute, one team would deliver two to three stones per day, so roughly 1,000 stones per day could be delivered. After the stone is delivered to the location where it is to be placed and trimmed or dressed, another team of laborers and masons use

Block with mortar, Khufu's pyramid, Giza.

levers and wooden rollers to move it into its final position on the row currently being constructed. A thin film of mortar would be used to act as a lubricant so the block would slide into position next to the adjacent block. Mortar was also used to fill in gaps between stones.

No great precision was required for the majority of the interior core stones—those placed to fill in the void between the perimeter courses and the internal structures—and the process was considerably simpler and faster. The stone was brought to its location on a sledge and simply rolled off and tumbled or levered into position. Stones of odd sizes and shapes were fitted in place expeditiously, as can be seen where the core stones are exposed in each of the Giza pyramids. Close joints were not required; the joints between these stones were filled with sand, rubble, or mortar to ensure that they remained in place. Careful leveling was not required at every course; instead, a series of benchmarks (say four as a minimum) could have been established. A 5- or 10-cubit-long pole was embedded at these benchmarks, and when the work reached a height corresponding to the end of the pole, a new benchmark would be established and the process repeated. Periodically the entire working surface was probably leveled and checked by some other means to maintain control. The logical point to do this would have been at Courses 9, 19, and other courses where a band of larger blocks was placed. The surveyors established vertical and horizontal control lines for checking alignment as the construction progressed.

We can see an actual example of these lines, thanks to some Mamluk tomb robbers who opened a huge gap in the north face of Menkaure's pyramid in an unsuccessful attempt to break in during the twelfth century. These remarkable photographs show marks that are 4,500 years old. Now, after exposure to the elements for 800 years, they are fading and someday will disappear.

As the project entered its third year, work was under way on a massive scale. Multiple faces had been worked on in the quarry. In each section, a small surveying crew laid out the area to be developed. They marked cutting guidelines for the quarry workers as they cut channels to free each block of stone. Their hardened copper chisels and stone adzes

Surveyor's red ocher lines at Menkaure's pyramid, Giza. ABOVE: Vertical control line, center axis. BELOW: Horizontal leveling line.

dulled rapidly, but an on-site workshop fabricated new tools and sharpened the old ones, providing a constant supply.

Excluding the large backing stones that are placed at Courses 1–10, 19, 35–38, 67, 90–91, 94–95, 98–100, and 118–19, the majority of the stones weighed between 1 and 2 tons. The labor to cut a stone this size is estimated as 1–2 labor-days. In Lehner's pyramid reconstruction project, twelve quarrymen using modern tools produced 186 blocks in twenty-two days, or an average of 1.4 labor-days per block.[2]

Once the stone was removed from the quarry face, it was moved to a staging area for finishing. Structural stones that were part of chambers or corridors were frequently left with notches, holes, or raised bosses that could be used to manipulate the stone into position. It is likely that little was done to stones installed in the interior of the pyramid other than to knock off any gross irregularities. Stones destined for use in or adjacent to the corridors, ceilings, walls, or other architectural features were dressed carefully to finished dimensions, as were the backing stones that made up the perimeter of each course.

When a stone was ready for placement, it was levered onto a wooden sledge and secured with rope. A team of laborers dragged the sledge from the quarry up to the construction ramp and then up to the level where it was placed. Depending on the size of the block, there were from 10 to 100 stone haulers. Hundreds of laborers were needed for the massive granite roof beams.

The process of moving heavy objects on wooden sledges is depicted in several tomb paintings. The most famous of these is in the Twelfth

Untrimmed stones showing lifting bosses, slope marks, and lever sockets.
TOP LEFT: Stones with lifting bosses.
TOP RIGHT: Lever holes. Note that the stones are precut to correct the angle of inclination on their edges.
BOTTOM LEFT: Stones partially trimmed with angle of inclination marked.
BOTTOM RIGHT: Lever socket in large stone.

Dynasty tomb of Djehutihotep, which shows 172 laborers moving a statue estimated to weigh 60 tons (three men per ton.)[3] One man is directing the operation, while another pours a liquid in front of the sledge. A similar process was probably employed at Giza.

The ramps were constructed of sand, soil, clay, and rubble placed between stone and mud-brick retaining walls. Vast piles of these materials were dumped into the ancient quarry when the ramps were demolished. Sections of the quarry, once mined of the best limestone, were undoubtedly backfilled with the spoils of the demolition of the ramps. The material (which is 10 to 15 meters thick) is a mixture of sand, soil,

Bas-relief from the tomb of Djehutihotep, showing 172 laborers hauling a 60-ton statue.

talfa, limestone chips, chips of other types of stone, and Fourth Dynasty pottery shards. Clearly, this is not undisturbed soil. The only logical reason for dumping such a large quantity of material into the quarry would be ramp demolition. Also, the easiest method of clearing the construction site would have been to simply push the ramp fill material over the edge and into the quarry.

Talfa, the local clay, has a low coefficient of friction when moistened, and tracks of talfa as wide as the sledges may have been placed on the ramps and periodically moistened to provide a smooth surface for the sledge runners. The sledge haulers walked outside this area, where it was dry and they had good traction.

The ramp had to be kept simple, because it was constantly being raised. It does not seem likely that the structure was ever very complicated. Whatever features were incorporated into the roadway were obliterated as it was raised to the next level. It does not seem plausible that the ancient Egyptians incorporated any features that had to be laboriously installed, then removed and installed again at each successive level as the ramp height increased.

An interesting concept for elevating blocks of stone was presented to me by a group of students at Lafayette College, where I spent several days as a visiting lecturer in 2000.[4] They postulated that a counter-weight system was used to move some of the heavier blocks. In this scheme, a weight first had to be positioned in an elevated location and connected by ropes to the load to be raised. As the weight was lowered, it helped raise the load. Several difficulties arise: First, the Egyptians had no pulleys, so they needed an alternate means of guiding the rope. Round stones with grooves called bearing stones may have been used for this purpose.[5] Second, this method requires raising the counterweight each time. The students proposed that the counterweight was actually a group of stone haulers, who, once they'd delivered their load, jumped on a plat-form and rode it back down. Their weight helped raise another load. Then the platform was hauled up empty and the process repeated. Several other researchers have considered this idea.[6]

During the early stages of construction, after the first few courses were in place, the pyramid resembled a large, flat stone pedestal. In line with the center of the north–south axis, but in from the east face by approximately 30 meters, two small walls of blocks marked the entrance to the Descending Corridor. Deep beneath the platform, a team of masons chipped away at the bedrock, driving the tunnel ever deeper at a constant angle of 26.5°. Baskets of debris emerged from the opening every few minutes, passed up by a human chain that reached down to where the masons labored. This material was dumped on a spoils pile, to be used later in filling voids in the rough cut interior stonework of the pyramid.

While this work went on, the pyramid continued to rise. When the structure had risen to Course 23 (an elevation of 40 cubits), it was ready for erection of the Queen's Chamber. To reach this point, the Ascending Corridor had been carefully framed in a gap left in the masonry as each course was placed. At this level, the Horizontal Corridor, which leads to the center of the pyramid, was constructed. Carefully dressed stones forming the side walls of this corridor were placed on the flat working surface, but the top of the corridor was left open for the time being.

Next, the walls of the Queen's Chamber were erected. The chamber is positioned directly in the center of the flat working surface, so its peaked roof is directly beneath the apex of the pyramid. The ceiling design slope appears to have been 5:3. The chamber is 11 cubits long and 10 cubits wide. The rough floor (never finished) is notched to provide a footing for the walls. The walls are 9 cubits high at the sides and 12 cubits high at the ridge line.[7] The stonemasons took great pains with the dimensioning and finishing of these stones, producing very fine work and tight joints. The walls were raised in panels 1.5 cubits high. There are six panels to the top of the east and west walls, and two additional sections to the apex of the roof. However, the panels are of different sizes, and only three of the joints are continuously level around the chamber. The finish is so fine, and the joints so even, that the joint is barely perceptible.

The long axis of the Queen's Chamber runs east–west. On the east end, the workers constructed a small alcove, 2 cubits deep and 3 cubits wide at the base, also constructed of finely dressed stones. (See Plate 9.) The first two courses are vertical. The third course overlaps the first two by about 6.5 centimeters (one-fourth of a cubit), as do the fourth, fifth, and sixth courses. At the top (elevation of 9 cubits), the width of 1 cubit is bridged by triangular beams providing the end support for the V-shaped ceiling, a miniature version of the corbelled arch that will be used in the Grand Gallery. The purpose of the alcove is unknown, but on the basis of what other tombs have revealed, one can speculate that it was intended to hold a statue of Khufu.[8]

A small "pyramid on the pyramid" was built to erect the Queen's Chamber. In other words, in the center of the flat working surface at Course 23, the first row of blocks was laid, forming the walls of the Queen's Chamber. Next, small auxiliary ramps were constructed to each side of the chamber to provide access for successive blocks to be brought up and placed on top of the first row. Workers inside the chamber helped place the blocks and ensured that the inner faces were lined up vertically and plumb. As the walls rose, additional interior blocks were placed around them for structural support and stability. Although carefully joined and fitted, these interim backing blocks were ultimately buried in

Hauling and placing stones.

the body of the pyramid and not dressed finely like the blocks of the chamber itself.

While the highly skilled stonemasons who specialized in chamber construction were doing their work, two other activities were in progress. Another specialized gang of stonemasons constructed the first segment of the Grand Gallery, using a similar approach. Additional gangs were busy placing the blocks that would eventually complete Course 23. Initially, they worked in other areas to give the chamber crew unhampered access. (At this level, the flat surface of the pyramid is a square 200 meters on a side, or approximately the area of four football fields.)

As the lower walls of the Queen's Chamber were completed, Course 24 was built up in this area to provide a surface from which the stonemasons could work. Once all the chamber walls were in place, the entrance was sealed temporarily. Laborers brought up baskets of sand and filled the chamber to the top of the walls. Then, starting at one end, the pointed saddle roof (V-ceiling) beams were placed. The ends of these slabs were mitered to the correct angle to form the ceiling. Temporary shoring blocks were placed on the sand fill to position the ceiling beams at the center, while wooden falsework, lubricated with water or oil, provided a platform from which the ceiling beams could be levered and pulled into position. Once in place, they were restrained until the interior backing blocks could be placed at the outer edge to hold them. The vertical load of the ceiling beams was resisted by the chamber walls, and the horizontal thrust was taken up by the interior backing stones and eventually by the successive horizontal courses of masonry. Once all the ceiling beams were in place, the chamber was emptied of sand, and the supporting wooden falsework was removed. Since the slope of the ceiling is 30.5°, the slant length of the ceiling beams is 3 meters. These beams extend beyond the wall face an added 3.1 meters, which means that their center of gravity is beyond the wall face and they are in effect cantilever supports rather than arch supports. This is proven by the existence of a small gap at the roof apex.[9] If the beam failed as a cantilever, it could still support load as an arch, so the design provides a degree of structural redundancy. Also, the roof beams may be doubled—that is, there may be a second set above the first, although this could not be known without excavation. It is likely, however, because doubling beams was a common Egyptian building practice at the time, and it can be seen in a similar pointed saddle roof over the entrance to the Descending Corridor.

Arnold suggests that a construction gap was left open while the Queen's Chamber, the Ascending Corridor, the Grand Gallery, and the King's Chamber were built to allow for construction of the chamber roof, and it could have served as an interior access ramp. He cites examples of construction gaps that have been seen in other structures.[10]

With no access to the area above the Queen's Chamber, we do not know how the vertical loads above the chamber roof are resisted. It may be that the span is such that the ceiling beams alone were sufficient. Or it may be that the space above the ceiling beams was bridged with granite beams to provide additional support, as above the King's Chamber.

At Course 23, the block height is 1.6 cubits (84 centimeters). Each block weighs 3.5 metric tons. The block weight at Courses 22–23 is increased to 3 to 4 metric tons to provide added stability for the Queen's Chamber and reinforce the overall structure—a size that can be handled by a crew of between eighteen and twenty-four men. Meanwhile, as work on the chamber progressed, a steady stream of laborers would bring

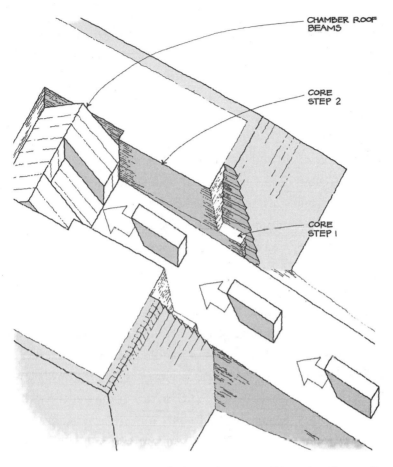

An internal ramp, possibly used to place roof beams.

blocks up on wooden sledges to the flat working surface. When they reached the proper position, the block was levered off of the sled onto a set of rollers, and the men returned to the quarry for another load. The stonemasons maneuvered the block into place, completed any rough trimming necessary, and then levered it into its final position, directed by an overseer. The outer perimeter backing stones were already in place, so the precision work was complete. The next step was filling in the void. Since Course 23 had already been accurately leveled, there was no concern about precise leveling for the next dozen or more courses.

The core blocks, installed next, were rough hewn. While close to the size of the backing stones, many were irregular. Any voids or gaps were filled with sand or rubble.

It took around 21,000 blocks to finish this course. Since the ramp enabled crews to place several thousand blocks per day, it took ten days to complete Course 23. From start to finish, the erection of the Queen's Chamber required approximately four and a half months.

Construction of the Grand Gallery proceeded by first laying a floor that rose at a slope of 2:1. While this appears to be the design intent, the as-built slope is slightly off, 26.2° rather than 26.5°. A channel 2 cubits wide and 1.16 cubits deep, with a 1-cubit-wide shelf or ledge on each side, is cut into the center of the floor, which is 4 cubits wide on the inside dimension. A series of notches is cut in the shelf at regular intervals on each side. The purpose of the notches is not known, but they and the channel may have secured the plugs that eventually were released to close the Ascending Corridor.

The first course of the Grand Gallery was laid on this sloping floor and somehow tied in structurally. The first two courses rise vertically for 2.29 meters (4 cubits, 10 digits). On top of these courses the masons then placed seven additional courses, each overlapping the one below by 1 palm in width (one-seventh of a cubit), so that at the top, the width of the gap is 2 cubits—the same as the channel in the floor. The vertical distance from the floor to the top of the gallery varies from 8.5 to 8.7 meters and averages 16.5 cubits.

The width of the wall blocks is not known, but to prevent their overturning as the structure was raised, they would have to be more than 2 cubits wide. To insure against the toppling of the gallery as the side walls were erected, wooden spacer beams may have been used to brace the top courses and keep the spacing true and constant. At the top, roof beams notched with a 2-cubit-wide projection completed the corbelled ceiling. The beams were cut into each succeeding side wall top course a few centimeters at the lower end, so they were restrained from sliding down toward the lower end of the gallery.[11] Before the gallery was enclosed, the granite blocking plugs were placed inside it and secured in an elevated position. They were slightly wider than the lower section of the Ascending Corridor so that when released, they would slide down by gravity and jam into the end of the corridor, closing it to access from below. This operation had to be performed in such a manner that the movement of personnel and ceremonial items through the Grand Gallery was not impeded when it came time to bury the pharaoh. It is not known exactly how this was done, but there is no doubt that it was: the plugs are there. Today, much of the detail just described is hidden by a wooden walkway that has been placed over the floor of the Grand Gallery for the convenience of visitors.

The Grand Gallery extends from Course 23 to Course 48, a rise of 21 meters (40 cubits). At its upper (south) end a large step rises about 89 centimeters (1.7 cubits) above the gallery floor. Heavy backing stones were placed at the lower end of the gallery to resist the downward thrust of the side walls. The side walls were erected in panels of sufficient length to extend upward through two to three courses of masonry to enable precise measurement and surveying, and for ease of placement from the flat surface of the current working surface.

When the construction reached Course 50 (43 meters or 82 cubits high), preparations began for building the King's Chamber. First, a short, horizontal passage leading from the end of the Grand Gallery to an antechamber and then to the entrance to the King's Chamber was laid out. A step up at the end of the gallery places the floor level of the antechamber and the King's Chamber at this elevation. The antechamber is 2.9 meters (5 cubits, 4 palms) long, 1.6 meters (3 cubits, 1 palm) wide, and 3.9 meters (7 cubits, 3 palms) high.[12] Within the antechamber are the slotted walls that house the portcullis.

The King's Chamber is 20 cubits long by 10 cubits wide and 11.33 cubits high. The walls are granite blocks, each course slightly irregular, and none of the horizontal joints (except possibly the first course) are closely leveled. Petrie reported that nineteen of the joints showed signs of displacement or opening; the walls are widened in some locations, and all the roof beams are cracked along the south side. He takes this as evidence of damage sustained during a major earthquake.[13] In some locations, the cracks were plastered over, indicating that some of the cracking occurred during construction.

The granite sarcophagus in the King's Chamber has overall dimensions of 2.28 meters long, 0.98 meters wide, and 1.05 meters high. Petrie observed that the sarcophagus is wider than the lower end of the ascending tunnel and therefore must have been placed in the King's Chamber while it was under construction and before the roof beams were placed. The sarcophagus has a number of imperfections, including dimensional variations, traces of saw cuts that went awry, and traces of a hole drilled to cut out the granite block used to fabricate it. These defects suggest that it was hurriedly done, possibly because it was late and delaying construction, or because it was a replacement for an earlier version that had been lost or damaged in transit from Aswan.

Why isn't the King's Chamber at the center of the pyramid, like the Queen's Chamber? Why is it offset by about 11 meters (21 cubits)? Why isn't it at an elevation of 80 cubits, rather than 82? While possibly insignificant, these discrepancies suggest that there was a design change, or that the builders made an error in interpreting the plans. One explanation is that the Ascending Corridor was longer than originally planned, perhaps to accommodate the granite blocking plugs. Another is that the antechamber and portcullis were afterthoughts and necessitated moving the King's Chamber.

Why was the King's Chamber 11.33 cubits high, when its other dimensions (and the Queen's Chamber) all measured in even cubits? August Mencken, an engineer, observes that the elevation of the ceiling of the King's Chamber is one-third (93.33 cubits) the height of the pyramid (280 cubits). With this number fixed, the floor of the King's Chamber would be at 82 cubits, and the *finished* floor of the Queen's Chamber (which was never installed) would have been at half this value, or 41 cubits.[14]

The roof of the King's Chamber consists of nine rough-polished granite beams. Their width varies from 1.15 to 1.59 meters and averages 1.31 meters (2.5 cubits). They are about 2 meters (4 cubits) thick. The entire width of the first and ninth beams cannot be observed because they overlap on the east and west walls. The visible portions are 57 and 59 centimeters, respectively, which suggests that the wall thickness is around 0.73 meters (or 1.4 to 1.5 cubits), assuming these two beams are of average width. Given that the chamber is 10 cubits wide, these beams must be at least 13 cubits, or nearly 6.8 meters long. At 2 meters thick, the weights range from 42.5 to 60 metric tons.

Placing these heavy beams posed a special challenge to the builders. They had the advantage of the main ramp to bring the beams directly up to Course 55. From this elevation, it was another 3 meters to the top of the King's Chamber. An auxiliary ramp to this level would have permitted the roof beams to be dragged into position once the chamber walls were in place and secured with backing stones. Like the Queen's Chamber, the King's Chamber was then filled with sand to provide a working platform, and timber falsework—supported by stones placed on the sand—supported the beams while they were maneuvered into position. The beams were most likely numbered and marked with positioning lines, as can be seen on the beams in the Relieving Chambers.

If the Grand Gallery had been kept open (or at least the upper end of it), teams of workers would have had enough space to assist in pulling the roof beams into position. In this case, the workers would be pulling downhill, so their body weight increased the effective pulling force.[15] They might have employed a system of counterweights in the Grand Gallery to assist in the effort, but I tend to believe that the Egyptians used simple methods to make the best use of abundant labor. The use of counterweights, although an interesting idea, would have needed some method of controlling the descent of the counterweight. In addition, the counterweight would first have to be hoisted to do any good, which seems like double the work.

The granite beams that make up the five Relieving Chambers are similar in size to the ceiling beams but are only rough-dressed. They are numbered and bear surveyors' marks indicating the center and end positions for alignment. Some bear graffiti naming the work gang that placed them, and a large cartouche on one of them bears Khufu's name. The beams in the first two chambers are supported on granite spacer blocks, 0.6 to 0.9 meters high, but in the third and fourth chambers, the spacers are limestone. The limestone spacers show evidence of distress and fractures—the result of the heavy compressive load they carry.

The Relieving Chamber beams may have been placed by extending the auxiliary ramp once the roof beams were in place. From the level of the main ramp to the height of the highest Relieving Chamber is a rise of about 17 meters. Since it is approximately 75 meters to the end of the main ramp, an internal auxiliary ramp could have been inclined 13°—steep, but possible. Such a ramp would have presented several difficulties. First, to move a beam weighing 50 metric tons would have required about 300 stone haulers, even with six lines—that is, 50 men per rope and a hauling line 60 meters long. Where would the stone haulers go? They would nearly fill the ramp, and unless there were some way for them to go down the other side, the pyramid surface would have had to be built up to accommodate them. Could they all fit in the Grand Gallery? Possibly, but any scheme to have them pull downward would have required some bearing surface to transfer the direction of force in the hauling ropes. H. John Hovland suggests that a greased log might have been used for this purpose.[16]

At the top of the Relieving Chambers (Course 81), the width of the pyramid is 133 meters. At this point, if the pyramid construction had been continued upward while the Relieving Chambers were built, there would have been a flat working surface extending approximately 67 meters from the center to the edge of the pyramid—the area available for crews pulling the beams up and maneuvering them into position.

A spiral ramp could also have been used to place these beams. The first segment, extending up to Course 91, would have permitted the beams to be brought up, with the added complication of having to make a 90° turn from the ramp onto the working surface rather than pulling them straight onto the working surface.

On the east and west sides of the Relieving Chambers, two massive limestone walls, entirely independent of the Relieving Chambers, provide vertical support for the masonry above. Differential settlement of these walls is evident in movement relative to the ceiling beams in the first and fourth chambers.[17] The beams are all roughly finished on the bottom side but totally undressed on the top side. They are of uneven sizes, many still showing horizontal and vertical alignment marks, some marked by small triangles. The marks are usually a broad line of red ochre paint superimposed by a fine black line to indicate the actual point.

In the fifth chamber, sloping roof beams made of limestone were installed as in the Queen's Chamber to make a pointed saddle roof. Holes in the floor of the fifth chamber seem to have been used to hold posts that supported the roof slabs during construction. These roof beams were also cantilevered. They have separated 2.5 to 3.8 centimeters at the apex, further evidence of differential settlement.

Once construction of the Relieving Chambers was complete, from Course 80 on up, the challenge lay in placing more blocks—the remaining 30 percent to be precise. At this point, all of the exacting work of constructing the inner chambers and corridors was complete. At Course 90 and again at Course 98, reinforcing bands of blocks 1 meter (2 cubits) high were placed. But from Course 100 to the top, the typical block height decreased to approximately 1 cubit, except for the periodic reinforcing bands at Courses 144, 164, 180, and 196.

Given the enormous difficulty of placing more than forty heavy granite beams halfway up in the pyramid—combined with the cracking and differential settlement that occurred even during construction—it is not surprising that this was the last time the pyramid builders placed the burial chamber high in the body of a pyramid. Hemiunu's pyramid builders had succeeded—perhaps better than they realized—but at a huge cost in time, labor, and materials. From this time on, the Egyptians reverted to placing the burial chambers on or in bedrock beneath the pyramid.

The Relieving Chambers raise questions. Why were they built? A pointed saddle roof, such as that of the Queen's Chamber, obviously worked. Why not do the same for the King's Chamber? (Possibly the antechamber would have interfered with the cantilevered roof beams.) Perhaps the cracks in the ceiling of the King's Chamber caused the builders to modify the design, and the Relieving Chambers were improvised to provide an added measure of safety. If so, why not place the pointed saddle beams above the flat ceiling with two chambers, or just enough to clear the antechamber? Why five chambers? These questions are reserved for future study. In any case, the pointed saddle roof became a standard in Fifth and Sixth Dynasty pyramids such as those of Userkaf, Sahure, Unas, Teti, Pepi I, and others who built at Abusir and Saqqara.[18]

The last segment of the spiral ramp enabled placement of blocks up to Course 194 (elevation 133.4 meters). From this point on, blocks were placed by working up from this level. A small auxiliary platform of stone constructed next to the end of the ramp would support a scaffold workers used to gain access to the top. As the last blocks were put in position, a "staircase" of stone was formed by leaving out some of the blocks and used to manhandle the last few blocks into place.

A rectangular notch was cut into the next-to-the-last course, consisting of four blocks, as a keyway to hold the pyramidion or capstone. The pyramidion, made of white Turah limestone, was raised into position and then grouted in place with a thin joint of mortar. Around the uppermost courses, casing stones—precut and dressed to the finish slope of 51.9°—were next installed. They were brought up the staircase to the proper level and slid along the shelf formed by that course until they were in the proper position. In the top twenty-five courses, minor deviations in alignment or slope would not have mattered, because they would have been imperceptible from the ground. Some minor trimming could have been done if needed. After the uppermost courses were completed, the staircase was filled in (with the aid of the scaffold) and the casing stones placed on it.

When all of the construction was completed down to the top end of the highest auxiliary ramp, the scaffold and auxiliary stone platform were dismantled. From this point on, precut casing stones were brought up the ramp on sledges and moved along the shelves formed by the backing stones until they were in position. The edges were trimmed on the ground to the exact angle of 51.9°, but excess material was left on the sloping face so the masons could finish the surface and make it uniform. Since all but a few of the casing stones have been removed, we cannot judge how successful they were in this effort. However, by examining Khafre's pyramid, the top third of which retains its casing stones, we can infer that the result was at least as good if not better on Khufu's pyramid.

For the top 100 courses, where the height of each course was on the order of 1 cubit, laborers could stand on the ledge of the course below (or even two courses below) to place the block and complete the final dressing and trimming.

Casing stones would be brought up the ramp and off-loaded onto the next ledge to be finished. If the ramp terminated at the south face, for example, casing stones would be brought to the end of the ramp, slid onto the ledge being worked on, and moved around the pyramid to the center of the opposite (north, in this case) face. Two teams could work in this manner, one moving blocks around clockwise, the other, counterclockwise.

This process did not have to be limited to the top level of the auxiliary ramp. Casing stones could be brought up and off-loaded at each course intersected by the ramp. Thus, work could be carried out on a number of courses simultaneously. However, at levels other than the end of the ramp, workers could move casing stones only counterclockwise around the pyramid, because the ramp would be in the way of the clockwise transfer of stones.

As the blocks were slid close to their final position, teams of masons fine-trimmed the edges and levered the block into position next to the last block installed. Then they dressed down the block to the angle of the face, continuously checked with wooden straightedges and levels calibrated to the pyramid slope. Periodically, the surveyors sighted back to benchmarks on the ground to ensure that the correct slope was being maintained.

The casing stones were half the size of a regular stone of a given height. On the top 100 courses, a typical weight was 400 to 600 kilograms. The smaller ones could be maneuvered by four to six laborers, while six to ten would suffice for the larger ones. Levers, rope harnesses, and rollers were used to position the casing stones.

While writing this book, I had an unplanned opportunity to check these numbers. With the help of three brawny men, I maneuvered a 400-kilogram photovoltaic power system into final position on the roof deck of my home without benefit of levers or rollers. In retrospect, my planning for this operation was seriously deficient, since we accomplished the job by brute force, with none of the sophistication employed by the ancient Egyptians.

After the casing stones were placed, the auxiliary ramp was demolished level by level to clear the shelf area previously taken up by the ramp, and the casing stones were filled in at the gaps on those levels. The work continued in this manner down to the level of the main ramp.

Some of the casing stones on Course 1 still remain on the north face of the pyramid. From the manner in which the platform stones beneath them are cut, it is apparent that these large casing stones were installed with excess material and then dressed to the final angle of inclination. This trimming process is clearly evident on Menkaure's pyramid.

The casing stones on the lower courses (say the first two or three courses) may have been placed by working from the bottom up, which would provide a reference point for back measurements to keep the angle of inclination constant. If so, at some level there would be an intersection between the casing stones being installed from top down and those being installed from the bottom up—most likely at Courses 20 and 21, two courses with smaller blocks. Here it would be easier to fit the blocks and make any necessary adjustments. The ancient Egyptians were clearly capable of inserting a casing stone between two completed courses, since it had to be done to close the entrance.

Installing the casing stones "bottom up" to Course 19 would have offered another advantage. All of the casing stones beneath the entrance could have been installed earlier. Once all of the casing stones had been installed on the lower portion of the north face, a ramp leading to the pyramid entrance on Course 19 could be constructed, and material removed as the main ramp was demolished could be recycled into the entrance ramp.

The bottom-up approach could not be used on the south face until the main ramp was demolished and cleared away from the face of the pyramid. Some means of raising the blocks to the various courses would have been necessary for this approach. Possibly the entrance ramp on the north side served a dual purpose, being raised by stages to install the lower casing stones, and then serving as the means to enter the completed pyramid.

Once the main ramp was demolished, the debris removed, and the site cleared, work could begin on the enclosure wall, courtyard, boat pits, and other structures that were part of the pyramid complex. Although the pyramid was the central and most visible component of Khufu's funerary complex, the other parts—the mortuary temple, causeway, and valley temple—were of enormous religious significance, since they were essential for maintaining the pharaoh's spirit in the afterlife. From the bare traces that survive, it is apparent that they were glorious structures in themselves. Sadly, we are unable to visualize them as they must have been—beautiful stone memorials to the king who built the greatest structure on earth.

Trimmed and untrimmed casing stones, Menkaure's pyramid, Giza.

CHAPTER 8

THE WORKFORCE

Among the many questions surrounding the construction of Khufu's pyramid, three in particular have intrigued humankind for centuries: How many workers did it require to build this immense structure? Did they work voluntarily, or were they slaves? How long did the construction take? In this chapter I will address the first two; in Chapter 9, the third.

In a sense, the questions are interrelated. Up to a certain point, the construction schedule depends on the available workforce. I say up to a certain point, because there is a point at which simply throwing more workers at a project does not hasten its completion. They may begin to get in each other's way, material production and delivery can't keep pace, operations slip out of sequence, or other logistical factors can interfere. Also, in any project there is an optimum workforce size for completing the job most efficiently—the right number of workers with the proper skills to complete the job in a systematic, organized manner.

Since the ancient Egyptians built three large pyramids (Meidum, the Bent Pyramid, and the Red Pyramid) in the fifty-year period prior to constructing Khufu's pyramid, it is reasonable to assume that they had evolved an orderly sequence of construction activities necessary to accomplish the project within the desired time frame. In a general sense, we know that the structure had to be completed within the pharaoh's lifetime to be ready before he died and entered the afterlife. According to the reigns of the eleven pharaohs who ruled during the Third and Fourth Dynasties, this could be as short as three or four years or as long as twenty to thirty years.

We do not know the exact nature of the workforce that built the pyramid; however, there is ample evidence that both laborers and highly skilled workmen were paid for and achieved prestige by the work they performed. Graffiti on some of the stone blocks indicate the names of the crew that cut the original stone or finished it. Administrators kept records of the receipt of materials, inspected them for quality, and arranged for paying workers and suppliers. Records were kept on clay tablets or papyrus.

One can marvel at the challenge of orchestrating the large workforce and making sure that it was housed and fed, that the right materials were present, that blocks of different dimensions and types were available when the time came to install them, and that shipments of materials by barge and boat down and across the Nile could be coordinated with the needs of the workforce. All of these tasks require a sophisticated project control and project administration system.

Workforce Size: Preliminary Assessment

There are four operations in which labor shortages could conceivably constrain the project's date of completion: cutting stones in the Giza quarry, transporting cut stones from the quarry to the pyramid course currently being constructed, constructing the ramps, and placing and trimming the stones. Analysis shows that all of the other operations (constructing roads and a harbor, leveling the site, placing the casing stones, etc.) are less labor intensive.

Quarry Operations

The quarries were required to produce 2.3 million blocks of stone of various sizes—2.2 million blocks of limestone from Giza, 98,000 casing stones from Turah, and several hundred granite beams and blocks (and the sarcophagus) from Aswan. Clearly, the 2.2 million blocks of Giza limestone drove the schedule. I assumed in my analysis that quarry workers cut a 20-centimeter-wide channel on three sides of a block, notched the bottom, and broke it loose. Two sides would be dressed in the quarry—the remainder, when the block was placed. Two labor-days were required to cut an "average" stone (say, 1.1 cubic meters) from the quarry. I compared this estimate with the data obtained by Lehner in the course of building a 6-meter-high pyramid. His crew of twelve quarry workers cut 186 stones by hand in twelve days, but they had the advantage of steel tools: equivalent to 264 labor-days, or 1.4 labor-days per stone. Given the limitations of the equipment available to the ancient Egyptians, 2 labor-days per stone on the average seems reasonable.

Khufu's reign is believed to have lasted for 23 years, from 2551 to 2528 BC. Assume that the construction project took twenty years and that each year had 280 workdays, or 5,600 workdays for the project. During this time, one quarry worker could produce 2,800 stones. Therefore, *on average*, to produce 2.3 million stones would require 821 quarry workers. In reality, many more stones were needed in the early phases of the project, so the optimum number of quarry workers might be three or four times this number. Also, if a shorter construction schedule was planned, the number would increase.

Granite is much harder than limestone and requires more time to cut and dress to final dimension. When I made a similar analysis for granite, I found that it took 100 labor-days per cubic meter to prepare granite blocks and beams.

BLOCK DELIVERY RATES AND TEAM SIZES

The size of hauling crews can be estimated in various ways. Assuming that a laborer could exert a lateral force of 350 newtons and that loads would be moved at 2 to 4 meters per minute, it would take 3.33 labor-days per block to bring blocks from the quarry to the construction site (roughly 300 to 350 meters). Carrying capacity ultimately depends on load and distance. Arnold reports that on a flat, well-lubricated surface, one worker can move a ton on a sledge. This figure drops to three workers per ton under practical conditions, and for a gradient of 9° (a 15.8-percent slope), about nine workers per ton.[1] Experiments showed that six workers could move a 6-ton load on a sledge on a flat, compacted, wetted earth surface.[2] Arnold also reports that fifteen workers could lift a 10-ton block with a harness and a single lever.[3] I assumed an average crew of twenty workers: fewer for the smaller blocks, and more for the large backing stones. Therefore, *on average*, to move 2.2 million blocks from the Giza quarry to the construction site over twenty years implies moving 393 blocks per workday, which takes a workforce of 1,300 laborers, or about 50 percent more stone haulers than quarry workers.

Moving blocks up the ramp to the working surface required more labor, depending on the size of the block, the slope of the ramp, and the distance the stone had to be moved. Estimating 10 labor-days per block on average yields 23 million labor-days of labor for moving blocks. Based on a schedule of twenty years and 5,600 workdays, *on average*, it would take 4,100 persons to move 2.3 million blocks up the ramps to the various levels of the pyramid.

EXCAVATION AND RAMP CONSTRUCTION

Laborers in ancient times moved dirt and sand in baskets. A load of around 0.028 cubic meters was typical.[4] The same methods are in use today, as I observed at the site where Hawass is excavating the workers' tombs. However, today the basket is made of synthetic material, rather than woven from reeds. The modern basket (see photo on page 141) has a useful volume of 0.025 cubic meters. For sand and soil with a density of 1.8 grams per cubic centimeter, the load would weigh 45 kilograms—the maximum a worker can handle, working continuously throughout the day. Tomb drawings show workers carrying sand and clay in baskets of about this size to make mud bricks. The remains of ancient ramps at several sites appear to have been constructed of rock retaining walls and filled with soil, sand, and debris.

On the excavation and the construction of ramps, I consulted nineteenth-century civil engineering data and established unit rates for moving earth manually.[5] These figures indicate that one worker can move 0.76 cubic meters of earth per hour with time added—depending on the distance the material was carried. Based on an average hauling distance for the size of the ramp, I estimated the labor rate at 10 labor-days per cubic meter for the excavation and placement of fill. Assume the ramps required placement of 1 million cubic meters of fill. This yields 10 million labor-days, or an average workforce of slightly fewer than 1,800 laborers over twenty years.

Keeping in mind that these figures do not reflect how the construction project would actually be staffed, and that they represent only a part of the labor requirements (but presumably the most labor-intensive components), the analysis can be summarized as follows:

· Quarry operations required 821 quarry workers over a twenty-year project schedule.

· Stone hauling from the quarry to the job site required 1,300 laborers.

· Moving stones up the ramps required 4,100 laborers.

· Ramp construction required 1,800 laborers.

In all, 8,021 workers are required to perform these four tasks working continuously over twenty years.

This estimate does not include many other tradesmen necessary to do the work, including carpenters; skilled stonemasons, who dress the stones to finished dimensions; sculptors, painters, and other artists; toolmakers; and surveyors, to name a few. If the number of workers is increased to 10,000 as an allowance for these other trades and skills, the total still does not include all of the support and administrative personnel necessary to supply, direct, feed, and house 10,000 workers—overseers, foremen, boatmen, cooks, butchers, bakers, and so on. If we use as a rule of thumb one support person for every man in the field, we'd have 20,000 workers.

The literature indicates that the construction occupied twenty to thirty years with a workforce variously stated as 100,000 to 400,000 people. Herodotus states, "Work went on in three monthly shifts, a hundred thousand men in a shift." Later he states, "It took ten years to build the track along which the blocks were hauled," and "to build the pyramid itself took twenty years."[6] These estimates seem high when one examines the work to be accomplished on a task-by-task basis.

What workforce could Egypt provide in 2550 BC? The population of Egypt at that time is believed to have been 1 to 2 million persons. Assuming it was only 1.5 million, and half of them were male, and half again of these were adults capable of working, the workforce is 375,000. If half of that number were farmers required to produce crops to feed the rest of the population, there remains a workforce of 137,500—certainly large enough to furnish the skilled craftsmen and trades necessary to build the pyramid. After all, the ancient Egyptians had already built three large pyramids in the fifty years preceding Khufu's ascent to the throne. There is also speculation that not all of the work was continuous, but that much of it took place when the Nile flooded and it was impossible for farmers to work in their fields. The time of flooding made available an additional large workforce of part-time laborers.

This analysis indicates that ancient Egypt had the resources to build Khufu's pyramid in twenty years—and most likely in a shorter time—if consideration is given to a logical sequencing and staging of the work to be performed. The workforce probably numbered in the tens of thousands, not in the hundreds of thousands.

WORKFORCE: DETAILED ANALYSIS

Based on the work breakdown structure described in Chapter 4, I prepared a much more detailed analysis of the workforce requirements.

First I considered stone block production requirements based on a five-, ten-, or fifteen-year construction schedule to see if the Giza quarry could produce the required amount of stone within this time. (Because of the smaller number of casing stones required from Turah and the longer lead time, casing stone production is not a limiting factor.) To meet these construction schedules, it is necessary to deliver 2.2 million stone blocks to the work area at these rates:

Block Delivery Rate

YEARS	BLOCKS/DAY	BLOCKS/HOUR
5 (1,400 workdays)	1,571	196
10 (2,800 workdays)	786	98
15 (4,200 workdays)	524	65

The blocks-per-hour number shown above is the *average number*, not the optimum. A more realistic production rate would be higher at the lower courses (when there are many blocks to place) and lower at the higher courses as the difficulty increases (and there are fewer blocks).

Using the five-year average number (196 stones per hour) to estimate the delivery rate at Courses 56 to 100, one can conclude that these delivery rates are indeed possible. This being the case, at lower levels the ramp would be wider and could sustain delivery rates twice this high. Above Course 75 the delivery rate drops off because of the lower number of blocks, so the constraint of ramp size and number of crews is eased.

If we assume that the quarry production rates could keep pace, or that block production could start in advance of the other work to stockpile blocks, the production of stone blocks for construction is not a limiting factor. There must have been ample materials on the Giza Plateau, since two more pyramids were constructed after Khufu's pyramid was completed. Based on these conclusions, the controlling factor in labor requirements is the construction of ramps (or any other method for delivering stone blocks to the working level of the pyramid, for that matter), as well as the stone delivery *rate*, which in turn is dictated by the size of the ramp.

The number of laborers required to haul each block can be established by considering the size of the blocks to be delivered to each course. This in turn fixes the delivery rates, because the ramps can accommodate only a certain number of teams at a single time, depending on the width and length of the ramp. From this point forward my estimate of the workforce will be based on this practical consideration, since it dictates, in turn, the number of quarrymen required to produce the blocks, the number of stonemasons to put them in place, and so on.

First, I considered the number of stonemasons required for the project. As indicated above, somewhere around 1,000 blocks of stone per day had to be installed. In reality, production rates were higher early in the project and then declined later as the number of blocks to be installed decreased. During the later stages of the project, a portion of the stonemasons would have been shifted to work on the mortuary temple, causeway, and other associated facilities.

To evaluate the size of the workforce, I assumed production rates of 1,000, 1,500, and 2,000 blocks per day and calculated the number of stonemasons and laborers required to cut stones, move them from the quarry to the plateau, move them up a ramp to the appropriate elevation, trim them, and put them in place. This calculation was performed for each course for both the internal core stones (2.2 million) and the 98,000 casing stones.

These calculations revealed that the required number of stonemasons at an average production rate of 1,000 stones per day varied from 2,250 to 7,000. These figures *do not* include laborers for hauling stones, which were considered separately. The higher number corresponds to the workforce needed to construct the first twelve and the higher courses with large stones, such as Courses 35–39, 47–49, and 90–91. Above Course 110, the number drops to 2,250 stonemasons.

It is unlikely that Hemiunu actually staffed the project with the high or low number of masons required. Instead, he would have picked a number in between these two extremes, which would cause production rates to be slower on the lower levels and faster on the higher courses. At the same time, stones could be cut in advance while the site was being cleared and leveled to create a stockpile for later use.

These options can be summarized as follows:

Number of Masons by Construction Rate

PRODUCTION RATES (stones per day)	CONSTRUCTION TIME (years)	MASONS REQUIRED (lower courses, large stones)	MASONS REQUIRED (higher courses, smaller stones)
1,000	7.9	7,000	2,250
1,500	5.2	10,500	3,375
2,000	3.9	14,000	4,500

I believe that around 4,000 masons were employed. The workforce would have been drawn from experienced stonemasons who had worked recently on the Red Pyramid or other building projects and apprentices recruited and trained for this purpose.

Next we can determine the population required to support a work-force of 4,000 stonemasons at the Giza site. Again, I am excluding the laborers who constructed ramps and hauled stones, many of whom may have been transient workers and did not reside permanently at the site, but were recruited from villages in the surrounding area.

UNIT LABOR ESTIMATES

I estimated 2 labor-days per block to rough-cut stones from the quarry at Giza and the white limestone casing stones at Turah. I further estimated 0.5 labor-days per block to dress the Giza stones down to finished dimensions, and 1 labor-day per stone for the Turah limestone, which had to be finished more accurately.

About 100 labor-days per cubic meter of stone produced were needed to rough-cut granite at Aswan, based on the requirement to cut or trim 0.3 cubic meters per cubic meter of finished stone product—the equivalent of a 20-centimeter channel on each side of the piece of granite being removed. This extremely laborious work was performed by hammering on the stone with round balls of another very hard stone (dolerite). For final dressing and finishing, I assumed it was necessary to remove 1.5 centimeters of material from each of five faces, requiring 25 labor-days per cubic meter.

Transporting Giza limestone from the quarry to the construction site took about 3.33 labor-days per block (average). Since the Turah limestone casing stones weighed less (1.5 tons average), I estimated that it took 1.67 labor-days for a crew of ten to move one of them from the quarry to the harbor; 0.13 labor-days per block for a boat crew of four to sail six stones across the Nile (boat capacity 10 tons), traveling 3 kilometers at 3 knots; and 1.67 labor-days to off-load and bring them up to the construction site: a total of 3.5 labor-days per block.

Similarly, for granite coming from Aswan, a barge with a displacement of 50 tons and a crew of ten could make the 900-kilometer trip downriver to Giza in 6.5 days, making 3–5 knots with a couple of stops, and 10 days to return. To load and off-load, I estimated 1 labor-day per ton each way, totaling 265 labor-days, or 5.3 labor-days per ton, or 16.4

labor-days per cubic meter. Fourteen labor-days per cubic meter were required to place granite beams and structural members. Arnold reports shipping records of a vessel that carried carrying six or seven blocks, a load of 15–20 tons.[7] The reported displacement of the boat unearthed at Giza is 45 tons.[8]

It took an estimated 32 labor-days per block (average) to move blocks up the ramps to Courses 1–3 and 18 labor-days per block for Courses 4–9. The larger stones at the lower levels obviously required larger teams but were not raised as high as the smaller blocks placed later on the higher courses. For Courses 10–25, 5.3 labor-days per block (on average) were required, and 5.4 labor-days per block for Courses 26–55. For example, in Courses 26–55, the average stone weight is 2,900 kilograms. To determine the number of stone haulers, I took into consideration the weight of the stone, the slope of the ramp, friction between the sledge and the ramp, and the force exerted by each laborer. (I assumed friction increased the required force by 50 percent.)[9] This calculation yielded a hauling team size of 20 laborers. On average, they can make 3.5 round trips per day from the laydown area to the center of the course in progress, pulling the sledge 3 meters per minute. The size of the ramp at this height is such that it can accommodate more than 500 teams, with ultimate capacity to deliver 1,953 stones per day. This requires a labor force of 11,160 stone haulers. I carried out a similar analysis for each section of the pyramid, estimating 2 labor-days per block for placing casing stones.

When the results were tabulated, I found that 10.4 million labor-days were required for hauling the core stones. It was possible to complete the placement of the core blocks in five years. The maximum number of stone haulers employed was 17,000 to 18,000 during the first year (the third year of the project), and the average number over the entire period was 7,500, not including other laborers who worked on site grading, road construction, and ramp construction.

I checked this calculation by computing the work theoretically required to place each block on each of 218 courses to the top, the product of force times distance: the distance is the height to the specified course, and the force is that required to lift a block of a given weight. This analysis showed that the total work done was 3,055 gigajoules. Allowing for rests and other stoppages, I estimated that a laborer can work at the rate of approximately 10 watts (1/75 horsepower), or 288 kilojoules per eight-hour day. This translates to 10.6 million labor-days, which compares favorably with the previous analysis.[10]

For constructing chambers and corridors, as well as removing the ramps, I estimated 2 labor-days per cubic meter for placing limestone beams and blocks, and twice that for granite. Removing ramps required 1.5 labor-days per cubic meter.

These unit labor estimates cover the majority of the labor-intensive direct construction tasks, but are by no means complete (see the table opposite). Other important trades and skills include surveyors, carpenters, brick makers, water haulers, basket makers, weavers, rope makers, metalworkers, sculptors, and artists. For example, the stonemasons used hardened copper chisels to cut blocks and dress the stones to finished dimensions. A team of toolmakers would be required just to keep the hundreds of stonemasons supplied with new or resharpened chisels. Lehner reports that one full-time tool sharpener had to be employed for every 100 stonemasons.[11] Carpenters were required to build and repair wooden sledges, manufacture levers and rollers, and erect scaffolding for stonemasons and sculptors working in confined spaces. Surveyors were continuously employed to keep the dimensions and angles of the pyramid true to plan as the structure rose.

Craftsmen were organized into guilds in which young men were apprenticed to skilled workers (possibly relatives) to learn the trade. No records from Khufu's time remain to show how these guilds were organized, and we have to make inferences from later periods. They probably consisted of foremen, supervisors, and at the top, a master carpenter or metalworker.

Workers were organized into crews and "gangs." We know this from graffiti found in hidden locations in many pyramids, including Khufu's and Menkaure's, where the names of the gangs are recorded for posterity. A typical labor crew might consist of 2,000 workers, in turn divided

Labor Estimates for Pyramid Construction Tasks

Task	Unit	Labor-Days per Unit
Rough-cut Giza limestone	Block[1]	2
Dress Giza limestone	Block[1]	0.5
Rough-cut Turah limestone	Block[2]	1
Dress Turah limestone	Block[2]	1
Rough-cut granite (Aswan)	Cubic meter	100
Finish granite (Aswan)	Cubic meter	25
Transport Giza limestone	Block[1]	3.33
Transport Turah limestone	Block[2]	3.5
Transport granite	Cubic meter	14.4
Site leveling	Cubic meter	2.5
Construct ramps	Cubic meter	3
Move core blocks up ramp:		
Courses 1–3	Block (average 14,000 kg)	32
Courses 4–9	Block (average 5,933 kg)	18
Courses 10–25	Block (average 2,600 kg)	5.3
Courses 26–55	Block (average 2,900 kg)	5.4
Courses 56–100	Block (average 1,700 kg)	3.2
Courses 91–146	Block (average 1,487 kg)	2.8
Courses 147–top	Block (average 1,000 kg)	2.9
Move casing stones up ramp	Block[2]	2.0 average
Place casing stones	Block[2]	0.5
Remove ramps	Cubic meter	1.5
Construct chambers/corridors; place limestone	Cubic meter	2
Construct chambers/corridors; place granite	Cubic meter	4

1. Data for an "average" block: 1.1 cubic meters, 2,400 kilograms.
2. Data for an "average" casing stone: 0.55 cubic meters, 1,200 kilograms.

into two gangs of 1,000 workers each. A gang was further subdivided into five groups of 200 workers each. As Hawass notes in the preface to this book, these groups are known as *phyles*, from the Greek word for "tribe." Each phyle had a name, like the crews who operated ships: Starboard, Port, Prow, and Stern. Each phyle was further subdivided into ten teams of twenty workers.[12] Based on this organization, one can speculate that each crew had an overseer; each gang, a supervisor; each phyle, a supervisor; and each team, a foreman: a supervisory staff of thirty-three persons per 2,000 workers.

In addition to the labor forces directly engaged in construction, a construction administration force included supervisors, warehousemen, and others responsible for logistics, including material suppliers, time-keepers, paymasters, clerks, and scribes. A small army of support personnel took care of the drinking water supply, food production and preparation, housing and shelter, first aid and medical services, port operations, ship transport, stevedoring, and other services. Priests prepared religious ceremonies and sanctified the work at critical stages. Teachers instructed the children of workers, many of whom were apprenticed and learned a trade. Soldiers or other representatives of the pharaoh provided security for valuable materials and royal property, resolved disputes, and maintained discipline and order.

Much of our recent knowledge concerning the workers who built the pyramids comes from excavations of the workers' tombs being undertaken by Hawass. His work, based on findings from hundreds of tombs, shows that the workforce was dedicated, skilled, and motivated. No slaves, these! In many cases, the tombs contain several generations, showing that son followed father as a skilled carpenter or stonemason. Pride in their work, as well as their sense of the importance of the project, is evident from the inscriptions found in the tombs.

I estimated the indirect or support labor force by developing ratios based on the number of construction workers. For example, I assumed one cook, one baker, and one brewer for every 200 workers (see the table opposite for other assumptions). The labor force at a large estate or villa as recorded in a Fifth Dynasty tomb at Giza was listed as two estate over-

CATEGORY	RATIO		NUMBER OF WORKERS
Stonemasons			4,000
Brickmakers	5/100	masons	200
Carpenters	1/100	masons	40
Foundrymen	1/100	masons	40
Tool sharpeners	1/100	masons	40
Surveyors	--		20
Rope makers	1/100	masons	40
Painters/artists	5/1,000	workers	22
Sculptors	5/1,000	workers	22
Subtotal (direct labor)			4,424
Foremen	20/2,000	workers	44
Supervisors	12/2,000	workers	27
Overseer	1/2,000	workers	2
Scribes and clerks	5/1,000	workers	22
Cooks, bakers, brewers	3/200	workers	66
Stevedores, warehouse workers	1/100	workers	44
Doctors and priests	1/100	workers	4
Security	5/1,000	workers	22
Subtotal (indirect labor)			231
Total direct and administrative labor			4,655

seers; a workforce supervisor; supervisors for the mess hall, linen, and other services; seven scribes; three butchers; two bakers; a cook; and others. The size of the villa is not stated, nor is the role and rank of its chief occupant, but these numbers suggest that my estimates of support workers are plausible.[13]

Total Labor Estimate

To determine the total labor required to build the pyramid, I divided the various tasks in the work breakdown structure into eighty activities. For each activity, I determined the quantity of work to be done, the number of labor-hours to perform the work, and the anticipated duration of each task. The workforce peaked in project years four and five, when the greatest number of laborers was needed for hauling stone. In project year one (2550 BC, or year two of Khufu's reign), the workforce consisted of a few dozen planners and designers. In year two, this number jumped to nearly 1,500 as the site work got under way, the Giza quarry began stockpiling stones, and preparations were made for building the main ramp. In year three, the quarry workforces increased eightfold, and the ramp construction was under way. Years five and six saw the bulk of the pyramid and its internal corridors and chambers completed, and then the workforce began to fall off as masons moved to other work, including the preparation of stone for the mortuary temple and the valley temple. When the full complement of 4,000 stonemasons and associated quarry workers and stone haulers was reached, slightly more than 400 carpenters, foundry workers, brick makers, rope makers, and tool sharpeners were supporting them. In addition, there was an administrative and program management group of 231 persons. (See table on page 215.)

Characteristics of the Workforce

The pharaoh was not the only one buried at Giza. Compelling evidence of the size and characteristics of the workforce comes from recent excavations of the workers' tombs.[14] Hawass is systematically excavating a large cemetery area south of the pyramids that consists of two parts. The lower portion is believed to contain the tombs of laborers; the upper portion contains more elaborate tombs that appear to be associated with craftsmen and overseers. So far, 43 of the upper tombs have been examined, and 630 tombs in the lower cemetery have been identified. Together they are believed to represent 20 percent of the tombs in the area, which would put the capacity of the cemetery at more than 3,000 tombs.

These tombs reveal a dedicated workforce of diverse trades and skills that worked on the projects at Giza. Some tombs contain three generations of workers, and wives and children are buried with husbands and fathers. Inscriptions in the tombs describe various categories of workers and supervisors and give their names. Inty-shedu was a carpenter who worked on boats; Nefer-theith's tomb is adorned with scenes of grain grinding and bread and beer making, suggesting that he was a bakery supervisor. Titles found in other tombs include "overseer of the side of the pyramid," "director of draftsmen," "director of workers," and "inspector of craftsmen."

Other excavations indicate a workers' village that may have housed thousands of people.[15] Lehner postulates that the walled city controlled access to the pyramid work areas through gates and roads leading to the massive entry gate in the Wall of the Crow. (See Plate 17.) The village had an eastern section, where individual structures housed the permanent workforce, and a western section, which housed a rotating labor force in barracklike structures. The complex also included bakeries; large, central silos for grain storage; a fish-processing facility; facilities for butchering beef, sheep, and goats; a commissary or mess hall; larger houses for foremen and overseers; warehouses; shops or storage facilities for such building materials as gypsum mortar, wood, and copper for tools; and workshops for craftsmen. (See Plate 18.)

The health of the workforce is also being studied as the tomb excavations continue. Giza workers had hard lives, and many of them died between the ages of thirty and thirty-five. The overseers and skilled craftsmen buried in the upper cemetery appear to have been healthier and to have lived five to ten years longer. Medical examination of the remains reveals herniated disks and degenerative arthritis, probably reflecting years of hard labor, in both men and women. Bone wear was also observed in knees.[16] Some skeletons also show fractures in the upper and lower limbs. Many of these injuries had completely healed and showed good realignment of the bones, indicating that medical treatment had been received. Even amputations had apparently been performed successfully. Teeth appeared to be generally healthy but were worn down—perhaps because of grit in food and grains. The graves suggest a high incidence of infant mortality, perhaps from childhood diseases or iron deficiencies. Even in the poorer tombs, children were found buried with simple toys, indicating the importance these people attached to the family.

I assumed that workers lived in houses of four sizes, with dimensions typical of those in Deir el-Medina and other ancient ruins. A basic three-room structure (24 square meters) was for a single worker or one with one dependent. A four-room structure (42 square meters) was provided for a worker with dependents. Foremen or supervisors had larger four-room houses (60 square meters), and overseers' houses encompassed 200 square meters. The space occupied by housing, based on these assumptions, is shown in the table oposite.

WORKER PAYMENT AND CONSTRUCTION COSTS

Workers were paid with grain (to make bread and beer), oil, other foods, and cloth.[17] Payments differed according to their skill and rank. Niuserre, a Fifth Dynasty king, built a sun temple at a site called Abu Ghurab. Not a true pyramid, it nevertheless had a lower temple, a causeway leading up to a large courtyard, an altar, storehouses, areas for animal sacrifices, and a tall obelisk raised on a pyramid-shaped base.[18] Inscriptions here describe the supplies provided annually to support the operation of the temple: 100,000 rations of bread, beer, and cakes; more than 1,000 oxen; 1,000 geese; and more, suggesting the magnitude of provisioning a large pyramid establishment.[19] In addition, a substantial barter economy allowed a worker to trade food or cloth for services, or a worker with one set of skills might perform work for another, who would return the favor by making something for him. Some "moonlighting" also went on: workers used their free time to work for third parties.

Evidence from later periods (the workers' village at Deir el-Medina) indicates that there was a well-developed job administration system. Records show who worked and for how long. Absences were noted, performance evaluated, and disputes settled by senior personnel or judges in matters regarded as serious. Absence from work due to illness, a death in

Worker Payment Estimate for a Workforce of 4,656 Persons

Category	Number of People	Beer in Jug	Subtotal	Loaf of Bread	Subtotal	Living Space (m²)	Total Space (m²)
Stonemasons	4,000	1	4,000	2	8,000	24	96,000
Brickmakers	200	1	200	2	400	24	4,800
Carpenters	40	1	40	2	80	24	960
Foundry workers	40	1	40	2	80	24	960
Tool sharpeners	40	1	40	2	80	24	960
Surveyors	20	3	60	6	120	24	480
Rope makers	40	1	40	2	80	24	960
Painters/artists	22	1	22	2	44	24	528
Sculptors	22	1	22	2	44	24	528
Subtotal (direct labor)	4,424						
Foremen	44	3	132	6	264	60	2,640
Supervisors	27	4	108	8	216	60	1,620
Overseer	2	8	16	16	32	200	400
Scribes and clerks	22	2	44	4	88	24	528
Cooks, bakers, brewers	66	2	132	4	264	24	1,584
Stevedores, warehouse workers	44	2	88	4	176	24	1,056
Doctors and priests	4	6	24	12	48	60	240
Security	22	2	44	4	88	24	528
Subtotal (indirect labor)	231						
Total workers	4,655						
Married workers (50%)	2,328						
Dependents (wives and children)	6,983	1	6,983	1	6,983	18	125,685
Single workers	2,328						
Total population	11,638		12,035		17,087		240,457
Annual consumption (280 days)			3,369,660		4,784,220		
Consumption (life of job)			15,163,470		21,528,990		

1. Four housing sizes (square meters): worker, 24; worker with dependents, 42; foreman/supervisor, 60; overseer, 200.
2. Annual consumption at peak staffing level.
3. Life of job consumption at 50 percent of peak staffing level for nine years.

the family, or personal matters—such as "brewing beer" or "working on my house"—were duly noted and excused.[20]

Information from the Middle Kingdom suggests that an overseer received the equivalent of eight jugs of beer and sixteen loaves of bread as a daily payment.[21] Workers of lesser status received proportionately less. In the table on page 219, I took the permanent workforce roster and applied a pro rata pay scale to estimate how much bread and beer (and therefore how much grain) was needed to support the permanent workforce. I added in dependents (wives and children), assuming that half of the workers were single and half had one or more dependents. Workers with dependents receive extra food rations. The table indicates that 3.4 million jugs of beer were required per year, along with 4.8 million loaves of bread. This implies that the bakeries were producing about 17,000 loaves of bread per day, or about one and a half loaves per resident. During the peak construction years, the food supply would have been two or three times this amount, but much of it would have been prepared in the workers' homes, not at the site.

There is no way to precisely estimate the cost of the project. However, I have attempted to express it in kilograms of grain, since this was the principal medium of payment. If beer requires 0.3 kilograms of grain per 2-liter jug, and bread 1.3 kilograms per 2-kilogram loaf, the total grain requirement for beer and bread is 7.26 million kilograms, or about 8 million liters dry volume.[22] This is equivalent to about 1.67 million *hekats* and, at a typical yield of 25 hekats per stat, requires a farm area of 67,000 stats— equivalent to a strip about 92 kilometers long and one kilometer wide on each side of the river. This is a small percentage of the arable land in the Nile Delta. Besides, we know that Egypt had the capacity to feed between 1 and 2 million persons at that time.

This estimate applies only to the permanent workforce residing at the site. During the peak construction years, when the worker population swelled with additional laborers, more food would have been imported from other areas. Also, workers at Turah, Aswan, and other places were paid for their work on the project. The total labor expended was on the order of 34 million labor-days over the ten-year period shown in the critical path schedule—about three times the amount contributed by the personnel listed in the table on page 215. Payments in kind would also have been made for imported wood, copper, and other supplies needed for the project, but most of the materials (limestone, granite, rope, etc.) were produced locally, and the only cost was for the labor involved in their extraction and manufacture.

Building Khufu's pyramid was costly and certainly impacted the Egyptian economy of that time. Yet it was a manageable expense and justified by the importance of the project—a tomb for the king, a tomb that all of humankind would someday regard as astonishing.

CHAPTER 9

THE CONSTRUCTION SCHEDULES

To best understand the approach taken in this critical path analysis, it is important to understand the underlying assumptions. Let's start with the assumption that the project needed to be completed during Khufu's reign of twenty-three years. Many researchers have taken this to mean that twenty-three years could be devoted to the pyramid's construction, but I believe this is far too long, for several reasons.

First, there was no assurance that the pharaoh would live for twenty-three years. A prudent planner—and we can be confident that Hemiunu was prudent—would have thought in terms of a shorter schedule to be sure the work was ready in advance of need. He also would want to enjoy during his lifetime the recognition that would accompany this splendid accomplishment.

Second, it would be incorrect to assume that twenty-three years were available for the pyramid construction alone, given what we know about the obvious engineering and construction requirements of such a project. The site is in the desert, and there was a clear need to develop facilities for transporting materials and building facilities to house, feed, and support a workforce of thousands of persons.

Third, after the pyramid was built, it took time to clear the site and complete the mortuary temple and other components of the entire funerary complex. These components were a necessary part of the facility, and a proper burial could not take place without them.

Finally, it is unlikely that construction began in the first year of Khufu's reign. No doubt he spent at least a year consolidating his grasp of governing the Upper and Lower Kingdoms, confirming the loyalty of his various ministers, district governors, and tax collectors, and replacing those whose loyalty was suspect.

Using this broad framework, we can estimate the time available for actual pyramid construction:

- Consolidation of reign: 1–2 years
- Design and planning: 1–2 years
- Site survey and preparation: 1–2 years
- Build pyramid: To be determined
- Remove ramps and clear site: 1–2 years
- Build mortuary temple, causeway, and valley temple: 3–4 years
- Total time, excluding pyramid: 7–12 years

With seven to twelve years needed for other project tasks, how much time could be allocated to the actual construction of the pyramid?

Khufu's father, Sneferu, reigned for twenty-four years, and some of his predecessors reigned for as few as five to ten years. We can assume that since Hemiunu was well aware of recent history and the mortality of his pharaoh, he would have planned to complete the entire project in fifteen to twenty years. With this goal in mind, he effectively afforded himself a 20 to 30 percent contingency in the event the work turned out to be more difficult than imagined or was slowed by natural disasters, labor shortages, bad weather, or delays in material deliveries. Every contractor uses this technique to protect himself or herself against schedule delays. In modern terminology, we refer to the extra time that is built into construction schedules as "float." The float time provides a contractor with a means of overcoming delays in one part of a project while ensuring the overall delivery date. With so much at stake, one can be certain that Hemiunu had a construction plan that incorporated ample float.

Based on these considerations, the schedule goal for pyramid construction (exclusive of site preparation, construction of the mortuary temple and valley temple, and other major auxiliary works) would have been in the range of five to eight years.

CRITICAL PATH SCHEDULE ANALYSIS

How long did it take to build the pyramid? Today, critical path scheduling—one of the tools used by modern program managers—can help answer that question.

The critical path schedule is an extraordinary tool for determining the time it will take to complete a complex project. It also specifies which tasks control the project completion date. These tasks are said to be "on the critical path." This method, which has not been applied previously to building Khufu's pyramid, is a superior forensic technique for estimating how long it took to build the pyramid.

To develop a critical path schedule, it is necessary to break the project down into a number of subtasks, or "activities," that correspond to each of the steps needed to complete the work. Predecessor activities and successor activities are assigned to each activity. Predecessors are those tasks or activities that must be implemented before a successor activity can take place. Successors are those tasks that logically follow a given task. The duration of the activity and the resources needed to complete it are also specified. The number of labor-hours needed for each activity are one resource that can be specified, enabling the computer program to calculate the total resources required by the project (in this case, the total labor) at any point in time. Then the schedule is said to be "resource loaded."

The method can be illustrated by a simple example. Building a small concrete block wall entails six activities:
· Purchase concrete blocks
· Mark location of wall on ground with stakes
· Excavate trench for foundation
· Place forms and reinforcing steel
· Place concrete foundation
· Place masonry courses

"Place concrete foundation" is an obvious predecessor to "place masonry courses," since masonry cannot be placed until the foundation concrete has been placed and has had time to cure. In this example, each of the six activities lies on the critical path, since a delay in any one of the activities delays the entire project.

Suppose that a schedule analysis is done for this project, using as a resource one bricklayer to complete the activity "place masonry courses." To complete the job more quickly, two bricklayers could be employed.

The critical path schedule analysis would show the effect of adding more resources—that is, a second bricklayer—on the overall schedule.

Next, suppose that the builder decides that concrete blocks are too expensive, so he or she is going to cast the blocks instead of purchasing them. This introduces a new activity, "cast concrete blocks." The plan is to make half of the necessary blocks in one weekend. On the following weekend the builder plans to excavate and place the foundation and, while the foundation is curing, cast the remainder of the concrete blocks so the project can be completed in a third weekend.

"Place masonry courses" now has two predecessor activities. The start date for placing masonry depends on the completion of "place concrete foundation" and "cast concrete blocks." The critical path could run through the activities leading up to "place concrete foundation" if they took longer than expected (suppose an underground waterline or a large boulder were encountered); or it could run through "cast concrete blocks," if it rained and this activity were delayed or concrete block production took longer than planned.

Let us suppose the latter was the case: A shortage of concrete blocks puts this activity on the critical path and delays the project. What can be done? Here, to improve the schedule—to catch up—the builder could get an assistant to cast blocks during the week, or he or she could purchase enough blocks to finish the job on time. Either action would remove "cast concrete blocks" from the critical path.

This simple example illustrates several key points of critical path schedule analysis, the most important being the effect of sequencing construction activities. In a complex construction project involving thousands of interrelated activities, each depending in turn on others, the critical path and controlling delay are not obvious. Once they are understood, however, the builder can take steps to change the way certain tasks are performed, thereby accelerating the schedule. When this is done, a whole new critical path can emerge, suggesting still other activities that can be modified to further improve the schedule.

As important as what is on the critical path is what is not on the critical path. A delay in completing those activities not on the critical path does not impact the completion date of the project, within certain limits. The schedule analysis will show how many days of delay in noncritical path activities can be tolerated before those activities fall on the critical path. Knowledge of this extra time—the float—enables the contractor to divert resources (usually labor) to other activities, including those on the critical path.

A CRITICAL PATH SCHEDULE FOR KHUFU'S PYRAMID

I prepared a critical path schedule for construction of the pyramid by subdividing the work into eighty activities, divided into fifteen major categories:

1. Mobilize project supervisory team
2. Prepare designs and plans
3. Site preparation at Giza
4. Giza quarry operations
5. Turah quarry operations
6. Aswan quarry operations
7. Construct harbor at Giza
8. Pyramid construction
9. Construct ramps
10. Construct Courses 1–218
11. Transport and place casing stones
12. Dismantle ramps
13. Support activities
14. Clear construction area and demobilize
15. Commence work on mortuary temple

For each activity, I specified predecessor and successor activities based on the logic diagram shown in Chapter 4. I also specified the duration of each activity and the resources required in terms of materials and unit labor hours to complete the task.

The analysis was performed using a software package called Primavera Project Planner, more commonly known as P3. It is widely used in the construction industry to prepare schedules for large public works projects.

An arbitrary date of January 1, 2550 BC, one year after the start of Khufu's reign, was chosen for the start of the project. I used an Egyptian calendar in which each month consists of three ten-day weeks. The five extra days each year were assumed to be feast days, not spent working. Eight workdays per week gives 288 potential workdays per year, reduced to 280 days to account for additional holidays and religious celebrations.

In the figure on pages 228-29 the critical path is indicated by the dark bars—the shaded bars are not on the critical path. The critical path runs through the project planning and design activities, as might be expected, and then follows the preconstruction activities (clearing and grubbing the site, cutting rock, shaping the limestone massif, and leveling the site) before construction begins. The Descending Corridor and Lower Chamber are not on the critical path because they were worked on parallel to the construction of the pyramid itself.

After the site was ready, the critical path shifts to the erection of Courses 1 to 23. Note that cutting of stones in the Giza quarry is not on the critical path because the quarry production rate, by virtue of an assumed early start and an initial stockpile of stones, coupled with an expanded workforce, is able to satisfy the demand for cut stones. Likewise, I assumed that the quarries at Turah and Aswan initiated production early in the schedule to have their products ready when needed. Once the harbor is ready and water conditions on the Nile River are suitable, their products can be shipped to the site and stored for later use.

The critical path runs through the Queen's Chamber, since it and the Horizontal Corridor, linking it to the Ascending Corridor, must be completed before the vertical construction of the pyramid can resume (the Queen's Chamber is the controlling activity). The Grand Gallery and the King's Chamber are not on the critical path, because they are worked on while the vertical construction continues to rise around them. The same can be said of the Relieving Chambers.

Once all of the courses are complete, installation of the casing stones falls on the critical path. With the long lead time available to the builders, the production and delivery of the white limestone casing stones from Turah is not a controlling delay. The final steps in the completion of the

Activity ID	Orig Dur	Labor Force	Total Man Days	Years (1-18)
Reign of Pharoah Khufu Begins				
0	0	0.00	0	Reign of Pharoah Khufu Begins
0.1	280	0.00	0	Pharoah Khufu Consolidates Power
Mobilize Project Supervisory Team				
1	8	1.00	8	Mobilize Project Supervisory Team
Prepare Designs and Plans				
2	289*	0.00	0	Prepare Designs and Plans
2.1	120	10.00	1,200	Preliminary Design of Pyramid
2.2	48	8.00	384	Site Selection and Consecration
2.3	48	20.00	960	Final Design
2.4	73	22.00	1,606	Refine/Finalize Project Scope of Work
2.5	73	22.00	1,606	Refine/Finalize Project Schedule
2.6	240	20.00	4,800	Recruit Workforce
2.7	72	40.00	2,880	Initiate Procurement of Raw Materials
2.8	72	40.00	2,880	Shipping: Acquire/Build Vessels
Site Preparation at Giza				
3	701*	0.00	0	Site Preparation at Giza
3.1	24	10.00	240	Rough Survey
3.2	127	400.00	50,800	Excavate to Bedrock (Clear and Grub Site)
3.3	120	1,000.00	120,000	Cut Rock
3.4	45	1,000.00	45,000	Shape Massif
3.5	22	2,045.00	44,990	Final Leveling
3.6	90	555.00	49,950	Construct Access Road - Quarry
3.7	120	416.00	49,920	Construct Access Road - Harbor
3.8	48	200.00	9,600	Construct Temporary Housing
3.9	17	200.00	3,400	Construct Workshops
3.10	19	200.00	3,800	Construct Bakery/Brewery/Other
3.11	24	33.00	792	Construct Brick Factories (4)
3.12	497	200.00	99,400	Construct Workers' Village
Giza Quarry Operations				
4	1,389*	0.00	0	Giza Quarry Operations
4.1	10	10.00	100	Survey - Layout Quarry Faces
4.2	30	10.00	300	Mobilize Quarry Work Force
4.3	1,179	4,000.00	4,716,000	Cutting Stones - Giza
4.4	18	25.00	450	Cut Blocks for Harbor Quay
4.5	295	4,000.00	1,180,000	Dress and Number Stones - Giza
4.6	778	10,000.00	7,780,000	Transport Blocks to Laydown Area
Turah Quarry Operations				
5	1,010*	0.00	0	Turah Quarry Operations
5.1	10	10.00	100	Mobilize Work Force
5.2	464	200.00	92,800	Cut Blocks
5.3	232	200.00	46,400	Dress and Number Stones - Turah
5.4	536	300.00	160,800	Load/Ship Blocks to Giza; Unload
Aswan Quarry Operations				
6	494*	0.00	0	Aswan Quarry Operations
6.1	10	10.00	100	Mobilize Work Force
6.2	400	500.00	200,000	Cut Granite for Lintels, Sarcophagus, Etc.
6.3	84	340.00	28,560	Load/Ship Granite to Giza; Unload
Construct Harbor at Giza				
7	113*	0.00	0	Construct Harbor at Giza
7.1	45	1,000.00	45,000	Excavate Harbor and Canal
7.2	75	50.00	3,750	Transport Stone to Harbor
7.3	38	50.00	1,900	Place and Trim Stones - Quay and Harbor Wall

Critical path schedule for the construction of Khufu's pyramid.

Activity ID	Orig Dur	Labor Force	Total Man Days	Activity Label
Pyramid Construction				
8	747*	0.00	0	Pyramid Construction
8.1	20	20.00	400	Survey Base-Establish Working Points & Benchmark
8.2	54	20.00	1,080	Construct Descending Corridor and Lower Chamber
8.3	7	80.00	560	Layout/Mark - Test Fit First Blocks
8.4	38	30.00	1,140	Construct Ascending Corridor
8.5	21	30.00	630	Construct Horizontal Corridor
8.6	28	16.00	448	Construct Queen's Chamber
8.7	90	40.00	3,600	Construct Grand Gallery
8.8	99	60.00	5,940	Construct King's Chamber
8.9	7	20.00	140	Transport - Set Sarcophagus
8.10	80	250.00	20,000	Construct Relieving Chambers
Construct Ramps				
9	1,010*	0.00	0	Construct Ramps
9.1	560	4,000.00	2,240,000	Main Ramp - Courses 1-55
9.2	194	114.00	22,116	Main Ramp Retaining Walls
9.3	450	200.00	90,000	Auxiliary Ramps
9.4	74	57.00	4,218	Auxiliary Ramp Retaining Walls
Construct Courses 1 to 218				
10	1,330*	0.00	0	Construct Courses 1 to 218
10.1	39	18,144.00	707,616	Transport Stones for Courses 1-3
10.2	58	17,136.00	993,888	Transport Stones for Courses 4-9
10.3	143	15,876.00	2,270,268	Transport Stones for Courses 10-23
10.4	286	11,160.00	3,191,760	Transport Stones for Courses 24-55
10.5	248	10,000.00	2,480,000	Stockpile Stones - Courses 56-218 at Top of Ramp
10.6	244	7,500.00	1,049,520	Transport Stones for Courses 56-90
10.7	212	3,070.00	650,840	Transport Stones for Courses 91-120
10.8	148	2,870.00	424,760	Transport Stones for Courses 121-146
10.9	71	2,380.00	168,980	Transport Stones for Courses 147-164
10.10	68	1,730.00	117,640	Transport Stones for Courses 165-183
10.11	33	1,000.00	33,000	Transport Stones for Courses 184-Top
10.12	706	3,000.00	2,118,000	Place and Trim Stones
Transport and Place Casing Stones				
11	120*	0.00	0	Transport and Place Casing Stones
11.1	13	286.00	3,718	Courses Top -184
11.2	9	710.00	6,390	Courses 183-165
11.3	11	684.00	7,524	Courses 164-147
11.4	10	1,278.00	12,780	Courses 146-121
11.5	14	1,106.00	15,484	Courses 120-91
11.6	23	906.00	20,838	Courses 90-56
11.7	21	1,436.00	30,156	Courses 55-24
11.8	13	1,450.00	18,850	Courses 23-10
11.9	4	2,175.00	8,700	Courses 9-4
11.10	2	2,460.00	4,920	Courses 3-1
Dismantle Ramps				
12	468*	0.00	0	Dismantle Ramps
12.1	225	200.00	45,000	Remove Auxiliary Ramps
12.2	348	3,000.00	1,044,000	Remove Main Ramp
Support Activities				
13	1,143*	0.00	0	Support Activities
13.1	150	30.00	4,500	Make Rope (4.2cm Diameter, Etc.)
13.2	475	30.00	14,250	Lumber (Sledges and Levers)
13.3	1,143	35.00	40,005	Make/Sharpen Copper Chisels and Tools
13.4	840	360.00	302,400	Mud Bricks - Housing, Workshops, Ramps
Clear Construction Area/Demobilize				
14	24	100.00	2,400	Clear Construction Area/Demobilize
Commence Work on Mortuary Temple				
15	0	0.00	0	Commence Work on Mortuary Temple

pyramid are removing the auxiliary ramps and main ramp. These are on the critical path. At this point—ten years into the schedule and eleven years into Khufu's reign—work can be initiated on the courtyard, boat pits, and the mortuary temple.

The "optimal" workforce for the project on an annual basis is shown opposite. Data for this figure are summarized from the P3 program, based on the resource loading associated with each activity. P3 makes the calculations based on "early start" and "late start" calendars. Early and late start workforces are not plotted in the figure; however, with the early start scenario, the labor force peaks at 43,000 workers in the early months of year five (2547 BC), stays high for the balance of the year, and then decreases to about 15,000 in year six and 3,000 in year seven. With the late start scenario, the workforce builds up more slowly, peaks at 39,000 workers in the middle of year five, and does not fall off as rapidly as with an early start.

The optimal scenario attempts to minimize the peak size of the workforce by compromising between the early and late start scenarios. The workforce peaks at 26,100 workers earlier, in year four (2548 BC), but stays at this value longer, until the end of year six (2546 BC). The need for workers is reduced by delaying the completion of activities that are not on the critical path. Approximately 5,000 of these workers are permanent residents at the site, and 21,000 are temporary laborers brought in to assist the quarry workers, move stones from the quarry to the laydown area, finish the harbor, canal, and access roads, and move stones up the ramps. At the peak, 85 percent of the laborers (18,000) are required for the latter task.

Summarizing, in the early start, late start, or optimal scenarios, the total labor expended on the project is 33.7 million labor-days. These scenarios imply that the maximum labor force was probably in the range of 30,000 to 40,000 workers. It could have been as low as 26,000 and most likely did not exceed 43,000.

During the later stages of the project, the workforce decreases. Most of those working on the pyramid itself did not leave the site but were transferred to other projects, including the mortuary temple, causeway,

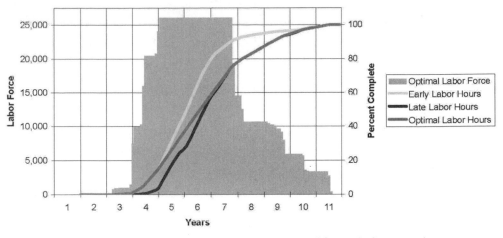

Construction workforce, Khufu's pyramid.

and valley temple, which now have need of their special skills. The completion of the pyramid marked an important milestone, but it was not the end of this project. Much remained to be done. With the site cleared and the ramp removed, the remaining work could be started. The courtyard surrounding the pyramid—nearly 20 cubits wide—was ready to receive limestone paving stones, and once that was complete, a wall, completely enclosing the pyramid, would be constructed. For the mortuary temple, on the east side of the pyramid, the planners envisioned a spectacularly beautiful structure with a floor of polished black basalt on which granite pillars would support the roof. The walls of the temple would be made of the finest limestone, carved in relief with religious scenes by the best sculptors and artists in the kingdom. The view from the mortuary temple would be spectacular, looking out over the Nile Valley, where the harbor and canal in the distance would reflect the morning sun. Then, from the mortuary temple, an elevated causeway would descend to the valley temple. Because of the steepness of the gradient from the plateau down to the valley floor, portions of the causeway would be elevated as high as 40 meters, and its length would be more than 740 meters.[1] The causeway walls would be graced with panels of magnificent relief carvings by the best artists, and the entire length would be covered with a stone roof.

Below, the valley temple would be constructed. Today we have only the barest hints of what it was like. It was far larger and grander than the temple built for Khufu's father at the Red Pyramid at Dahshur. It had a floor of black basalt. Other details are still concealed beneath the modern village of Nazlet el Simman, but archeologists have been able to speculate on its nature from the well-preserved ruins of Khafre's valley temple, and to a lesser extent from the shattered remains of Menkaure's valley temple. Khafre's valley temple was on the order of 45 meters square.[2] The walls were built of large limestone blocks and faced with granite brought downriver from Aswan. The structure had alabaster floors and a roof constructed of granite beams supported by granite columns. We can surmise that Khufu's valley temple was equally impressive.

At the site of Khufu's pyramid, additional work included the construction of the Queens' Pyramids and storage pits for the pharaoh's boats. Two large pits had to be excavated in bedrock on the south side of the pyramid. One of them, uncovered in 1954 and found to be 31 meters long and 2.6 meters wide, was roofed over with approximately forty heavy limestone beams, each weighing between 10 and 20 tons.[3] It was not large enough to hold the 43-meter-long boat with its 5.9-meter beam intact, so the boat had to be disassembled before it could be placed in the pit. As soon as the last vestiges of the main ramp had been cleared away, this work could begin. The Queens' Pyramids were constructed at exactly one-fifth the scale of Khufu's pyramid. Each contained 20,500 cubic meters of stone, or slightly less than 1 percent of the volume of Khufu's pyramid.

In summary, the major elements of the *as-built* schedule were completed as follows:

- Project planning and design: 12 months
- Site work and preconstruction: 30 months
- Pyramid construction: 80 months
- Remove ramps and site cleanup: 20 months

So far the project had taken ten years to complete, not including the courtyard, boat pits, Queens' Pyramids, mortuary temple, causeway, and valley temple. The as-built performance turned out to be consistent with

Hemiunu's original goal. It was now 2541 BC. No one could have foreseen that Khufu would reign for another thirteen years, as it is believed. During this time he would have seen the mortuary temple completed, the causeway built, and his spectacular valley temple finished. His boat pits were made ready but stood empty until his death, so he could continue to use his boat. Dismantling it would have to come later.

Although we are still uncertain of just how the Egyptians built the Great Pyramid, it continues to stand today, awesome testimony to the skill and sheer determination of the ancient race who built it. We also must stand in awe of their program management techniques. They must have had highly developed administrative and planning techniques, because the complexity and logistical requirements of the pyramid project are so great that it would not be possible for a single individual, or even an uncoordinated group of individuals, to carry out such an undertaking. The fact that the Egyptians could plan, organize, and execute an undertaking so complex, with the marshaling of so much labor—a significant fraction of their population—points to their remarkable skills in managing the work. Their program management was an accomplishment no less impressive than the legacy of stonemasonry they left behind.

CHAPTER 10

LIFE EVERLASTING

Once his royal residence had been built at Giza, it seems certain that Khufu visited the pyramid site from time to time during construction. He watched as the flat top of the pyramid grew steadily smaller, until finally the pyramidion was installed and the structure culminated in a perfect point. Then, as if someone had poured a giant bucket of paint on the top, slowly the pyramid became white—first the uppermost courses, then lower, as the workers installed the white Turah limestone casing stones. As the white casing grew ever larger, Khufu could have seen a diagonal scar slowly disappear from the face of the pyramid as the workers removed one of the upper ramps as they worked their way down. "Almost done," he thought, before remembering the entrance, the dark wound in the side of the otherwise perfect north face of the pyramid. It remained to seal the entrance and remove the north ramp before his perfect pyramid could be considered finished. Did he then experience a flash of insight, the soul-grinding reality, that completion required his death? It was an eerie feeling, knowing that the final steps could not be taken until he died, that his life was the final impediment to completing the project.[1]

To the best of our knowledge Hemiunu and his construction teams did finish all of the works required by Khufu. Where once there had been open desert east of the pyramid, a large walled city housing workshops, storerooms, workers' houses, granaries, and the studios of artists and sculptors emerged. A royal residence was built, and we can assume that Khufu resided there from time to time, near the walled city and the towering white pyramid that Hemiunu and his legions of construction workers had built.[2] We will never know with certainty the true extent of the pyramid city. Over the ages, Giza served as a source of building materials for expanding Cairo and the surrounding area. The site was systematically stripped of the ancient building stones. Climate changes affected it;

desert sands covered parts of it; the course of the Nile River gradually shifted more to the east, and new villages were constructed on top of the windswept ruins. Today, the likely sites of Khufu's valley temple and portions of the pyramid city are buried beneath the foundations of the modern village of Nazlet el Simman.

Despite these obstacles, tantalizing traces of the city have been found by archaeologists. Recent work suggests that the pyramid city continued to grow following the reign of Khufu until it could service the pyramids of his son Khafre and grandson Menkaure. On page 236 is an artist's concept of what the Giza site might have looked like as the construction of Khufu's pyramid neared completion—say, around 2541 BC. The beginnings of the pyramid city can be seen, along with the harbor, canal, quarry, and access roads. It is likely that these facilities were expanded later, when Khafre began construction of the second pyramid at Giza.

During Khufu's reign, Hemiunu as vizier was responsible for the day-to-day administration of the city. Scribes kept records of purchases and deliveries of produce, livestock, and raw materials. Inventories of goods stored in the king's warehouses were maintained. There were lists of priests on duty, artists working on projects, payments due workers. All of this came under his purview.[3]

Southeast of the great wall that enclosed the pyramid, the beginnings of a cemetery could be seen. Khufu had granted permission for a small plot of land to be set aside. Here, in the shadow of their greatest accomplishment, some of the foremen and craftsmen who had labored on the pyramid were buried.[4] On the west side of the pyramid, behind the great wall, the planners laid out a grid of narrow streets for another cemetery. Here mastabas were constructed for nobles and their families, including one for Hemiunu. Those who had served the pharaoh had the good fortune to be buried close by in the event that he needed their services in the afterlife.

As the work wound down, we can imagine that Hemiunu no longer spent every day preoccupied with construction matters, but rather turned his attention more and more to the state of the Egyptian economy. In some years the flow of the river was less than normal, and the

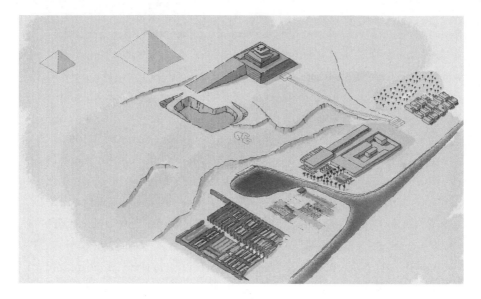

How the site may have looked as Khufu's pyramid neared completion.

crops suffered. The royal treasuries needed to be bolstered after the heavy expenses of the peak years of construction, but in general the economy remained vibrant and strong.

Khufu's attention no doubt was focused almost entirely on matters of state. He was deeply embroiled in keeping the district governors in line and arbitrating the endless disputes and jurisdictional issues that arose. His attention also turned to countries beyond Egypt's borders. There was a demand for Egyptian products: The fine rope made by Egypt was found on nearly every oceangoing vessel in the Mediterranean, and Egypt had a monopoly on the production of papyrus paper, which was widely sought outside the country's borders. Khufu was intrigued with the treasures his traders brought back from their excursions: metals, semiprecious stones, cloth, wild animals, and other novelties. They described the places they visited, and he concluded that none were equal to the two kingdoms of Egypt. It was a good time for Egypt, with no hint of foreign invasion, the country at peace internally, and the economy growing and robust. In return for labor, food for the construction forces, and raw materials, Khufu had made concessions of the royal lands to a number of nobles and religious leaders throughout the country. Although these land grants reduced the pharaoh's revenue somewhat, they had been necessary to support the building program. In theory, they were revocable at the will of the king, but as time passed, they passed from father to son and came to be considered a hereditary right. Khufu expressed no concern about this, but Hemiunu foresaw a time when the rising wealth of the temples and rich landholders would reduce the authority of the pharaoh—or worse, lead to conflicts with the court.[5]

Hemiunu posed for a statue, presumably to comply with Khufu's wishes. Hemiunu would have been filled with impatience as artists sketched him. The sculptor, hovering in the background, whispered directions to the artists, suggesting shading and small details that would guide his work. An assistant made measurements with a cubit rod to determine the size of the block of stone to be used for the work. The statue was carved from fine limestone, the inlaid eyes made of semi-precious stones—perhaps black onyx. It showed Hemiunu seated, wearing his robe. When the statue was finally completed and placed in the courtyard, Hemiunu must have been pleased with how he had been portrayed. Larger than life size (3 cubits high), it revealed a composed, dignified, and serene person.[6]

Hemiunu was also concerned with his own tomb. We have no way of knowing if, in spare moments, he sketched the outline of a mastaba himself or approved a design made by others. In either case, he knew that the space he'd been allocated was two or three times the standard. If tradition were observed, the interior of the tomb would be limestone, with finishes of alabaster or other fine stone. It would be decorated with scenes reflecting his various duties as vizier. There would have been a place for his statue. Did he monitor the work as he did the construction of Khufu's pyramid? If so, perhaps he better understood the complex thoughts that must have occupied Khufu's mind when the pyramid was being constructed—wanting and yet not wanting to see it completed.[7]

The date is now 2528 BC. Along with other members of the court, Hemiunu watched as the entourage of solemn priests exited the pyramid. As they came down the ramp, a dull sound echoed from within. Workers had released the first of the three granite-plug blocks that would seal the Ascending Corridor forever.

Months had passed since Khufu's death. His body, attended by his grieving family, had been brought to the valley temple, where it lay in state for the requisite period of time. Then it was brought to the mortuary temple, where it underwent ritualistic cleaning and purification. According to ritual, the pharaoh's remains were wrapped in a great length of finest linen and shrouded in a life-size shell of cartonnage painted in his likeness.[8] The mummy was then placed in a coffin of carved cedar and sealed. Khufu's palanquin, staff, flail of office, and a number of other personal effects were also carefully placed in the chamber. When all was in readiness, the coffin was carried into the pyramid, up through the Grand Gallery, and placed in the red granite sarcophagus. After the priests and attendants departed, the heavy lid was slid from its supporting framework onto the sarcophagus and locked in place to close it.[9]

We can imagine at this point how Hemiunu must have felt. So much work, so many years, and now it was done. As he left the ceremony to return to his residence, he must have thought about Khufu. He hoped that the pharaoh's soul was at rest, that his needs for the afterlife were fully satisfied, and that now, with this structure, his spirit could ascend the stair steps to the gods.

Later, Hemiunu saw to the sealing of the external entrance to the pyramid. The best stonemasons were careful that no trace of the opening remained. He inspected the work personally to see to its perfection. Then the funeral ramp was removed, and the site once again cleaned up.[10]

After paying his last respects, Hemiunu took one last look at the pyramid before leaving. In the coming days he met with the new pharaoh, Khufu's son Djedefre, who assumed responsibility for completing his father's projects. There were the matters of getting Khufu's boats into the boat pits and work still to be done on the valley temple.

When Djedefre ascended the throne, he decided to build a pyramid at Abu Roash, rather than at Giza. There is no record of why this decision was made. The site he selected is 5 kilometers north of Giza, and it occupies the top of a hill. From this vantage point, the pyramids at Giza can be seen in the distance. Little remains of Djedefre's pyramid; it is essentially a pile of rubble. Based on measurements of some of the foundation blocks and other debris, it appears that he planned to build a pyramid about two-thirds the size of his father's. The base is estimated to have been 200 cubits (106 meters) long, and the height was intended to be 110 to 130 cubits (57 to 67 meters).[11] From what can be observed of the construction methods, it appears that he returned to some of the techniques used in the Step Pyramid. The burial chamber was constructed by digging a huge pit in the ground and then roofing it over with a descending corridor for access. One unique feature of Djedefre's pyramid is a long causeway, over 1,700 meters long.[12] Djedefre's pyramid was never completed. The remains suggest that perhaps twenty courses were erected. His reign is believed to have lasted only eight years. Djedefre did not fail in his filial duties to Khufu, however. His cartouche appears on the limestone beams that conceal Khufu's boat pit, showing that he probably took responsibility for burying his father.[13]

Hemiunu may have served three pharaohs with dedication: Khufu, his son Djedefre, and, as an elder statesman, Khafre (Khafre's vizier was Ankhaf, one of Djedefre's sons). Hemiunu's heart must have been gladdened when Khafre started the second pyramid at Giza, but Djedefre's tomb at Abu Roash was probably an embarrassment to him.

Some years later, another notable funeral took place in the western cemetery. A large funeral cortege brought Hemiunu's coffin into his mastaba, behind Khufu's pyramid. The statue, which once had stood in the courtyard surrounding the pyramid, had been moved inside his tomb. Hemiunu's tomb was one of the largest and finest in the western cemetery, even more elaborate than the tombs of some princes and princesses. During the late afternoon on that somber day, the sun lit up the west face of the pyramid, making a dazzling white triangle. Behind that triangle, Khufu slept in the King's Chamber. In the shadows at its base, Hemiunu finally found rest, his great labor done.

With the completion of Khufu's pyramid, the construction of enormous stone monuments in Egypt had reached its peak. Whether through lack of will or failed planning, or out of respect, his son's pyramid, at 274 cubits, did not match the height of Khufu's Great Pyramid. From that point on, most pyramids built were 30 to 40 percent as high as Khufu's; none exceeded 55 percent of its height, and none achieved the same degree of engineering design perfection and grandeur. Thus, in less than 150 years, the era of building mammoth stone structures arose and faded away.

For some time—decades, possibly centuries—Khufu's, Khafre's, and Menkaure's retainers maintained the temples and the traditions that supported the cults of the Giza pharaohs. This support eventually waned, and the pyramids were targeted by unscrupulous individuals for the riches they were thought to contain.

Ironically, there came a time when the grandeur of the pyramids worked against them. Their size and prominence were advertisements to grave robbers, who sought the treasures believed to be contained within. The thieves' passages into the pyramids and mastabas showed uncanny accuracy as they bypassed the ancient designers' barricades—almost as if the thieves had inside knowledge of the openings, portcullis, and plugs that barred access. Once inside, they broke open the sarcophagus, ripped apart mummies, stripped them of gold and jewelry, and scattered the remains. Sometimes they left behind their grave-robbing tools, suggesting they had departed in haste.

Today, Khufu's pyramid stands somewhat diminished, robbed of the contents it was built to protect, stripped of its outer casing of white limestone, its supporting temples razed to the ground and their foundations buried. What of the future for this glorious monument? Having survived 4,500 years, what will remain for visitors forty-five centuries from now?

Obviously, this question cannot be answered with certainty. Tourists have visited the monuments for centuries, but in recent times their number has increased. Over the centuries, the footsteps of visitors, camels, and horses have taken their toll, as have vibrations caused by automobiles and buses. Increased humidity from the breath of visitors confined in underground spaces has damaged priceless paintings and even threatens the limestone itself. Modern construction continues to encroach upon and bury ancient ruins. The area where Mark Lehner has excavated bak-

Thief's wooden mallet propping up sarcophagus lid, mastaba at Meidum.

eries and food-process-
ing facilities is used by
riding stables. During the
winter digging season, it
is uncovered by Mark and
his team, and when the
weather warms and the archaeologists leave, the entire site is once again
covered up to protect it. Every year the site must be excavated and covered up again.

Many questions remain. We would like to know more about the
means and methods of construction. An extensive (and expensive) clearing of rubble on the south side of the pyramid would most likely reveal
traces of the ramp used to build it, but that would require disrupting
underground utilities and the modern access road. Likewise, further
excavation east of the Sphinx might confirm the existence of the harbor
and shed light on its size. Any sailor knows that objects fall off of ships
and sink in harbors with regularity, so a harbor excavation might yield
intriguing artifacts. If access could be gained to the area above the
Queen's Chamber, it would be interesting to see if there is a second set
of roof beams, as there is over the pyramid entrance.

Further study and measurements in the Relieving Chambers above
the King's Chamber might shed light on why they were constructed.
Why didn't the builders use the pointed saddle roof design that was
employed successfully in the Queen's Chamber? (It seems that this
design became the standard for most Fifth and Sixth Dynasty pyramids.)
Why were *five* Relieving Chambers built? Why was the King's Chamber
offset from the central axis of the pyramid? Finally, I hope that future
work at Giza will discover additional Fourth Dynasty tools, which could
shed light on construction methods.

A better understanding of the exact placement of stones in the interior of the pyramid, adjacent to corridors and chambers, for example, would reveal much about building methods, and therefore the time and labor required to place the stones. It is almost a certainty that ancient surveyors' marks still exist within the core masonry, as well as stone numbers and possibly more graffiti and dates. Unfortunately, at present there is no nondestructive method of collecting this type of information, and it is questionable whether one would want to cut an access tunnel into the interior of the pyramid for this purpose.

This brings forth another issue: exploration versus preservation. It seems there is a sort of Heisenberg Uncertainty Principle at work.[14] We would like to find answers to the many remaining questions, but the act of exploration itself modifies the monuments. Even the most careful work leads to some damage, and future exposure to weather and visitors accelerates the disintegration. How is this dilemma to be resolved?

Of great importance is careful management of archaeological sites and the entire process of exploration.[15] Meticulous record keeping, such as practiced now under Hawass's leadership, is also important. New technologies will help, and geographical information systems will capture the voluminous data that results.[16] Passive, noninvasive measurement techniques can guide fieldwork and minimize the risk of collateral damage, including the use of space satellites for accurate mapping, surveys using various wave-length sensors, radar, cesium magnetometry, and aerial photography.[17] Miniature, remote-controlled cameras have been employed at Giza with good results.

The problems of research are compounded by the importance of the monuments to the Egyptian economy. In discussions with me, Zahi Hawass outlined innovative changes to protect the pyramids. The Fayum Road is being reconstructed to keep traffic away from the monuments; a visitor's center and new museum will be built west of the pyramids; and walking paths and shuttles for transporting passengers will minimize the environmental impacts on the site. Implementing these changes will require the concurrence of local and national political leaders, reconciling the interests of diverse stakeholders, including merchants and taxi drivers, and most importantly, the acquisition of funding. On a larger scale, Hawass champions a broader role for Egyptian archaeologists and experts, and seeks the return of Egypt's treasures from some of the foreign lands in which they currently reside. This is a more difficult challenge, but one that merits understanding and cooperation.

Solutions to these problems must be found to protect and preserve the one remaining example of the Seven Wonders of the Ancient World. This structure—built as the stair steps to the gods of the ancient Egyptians—is an enduring monument to human creativity and ingenuity, and Egypt deserves international support to make certain that it continues to encourage and inspire future generations of designers and builders.

APPENDIXES

EGYPTIAN GODS

The Egyptians paid homage to many gods, and the preeminent gods changed over the millennia. In addition to the nationally recognized gods, there were deities unique to every great city, and even small cities or wealthy families may have had their special gods. Some gods rose or declined in popularity, depending on wars or calamities. The spelling of some names varies, depending on how one translates the sound into English. Below I list some of the better-known names.

AMON (AMEN) A local god of Thebes, usually shown as a human.

ANUBIS The god of mummification. Anubis is usually depicted with the head of a jackal. He assisted in the rites by which the deceased entered (or was denied entry) into paradise. He is usually shown holding the divine scepter carried by kings and gods.

ATUM (TEMU) The god-creator, an early form of Ra.

GEB (SEB) The son of the god Shu, sometimes called "the father of the gods." He became the god of the dead.

HATHOR The goddess of love, depicted as a female figure with cow horns. The deity of happiness, music, and dance.

HORUS Son of Osiris and Isis, he later hunted down and killed his uncle Seth to revenge his father's death. Usually depicted as a falcon-headed god. Considered the god of the day. In one hand he holds the ankh, the symbol of life, and in the other, the divine scepter carried by kings and gods.

ISIS   Wife and sister of Osiris. She had great magical powers.
Isis brought Osiris back to life after he was murdered by Seth. She had a
son, Horus, and was considered the protector of children.

MAAT   The goddess of truth. She presides over the weighing of
the heart when the deceased seeks entry to paradise.

NEPHTHYS   The sister of Isis, wife of Seth, was considered the goddess of women. She was associated with the home of Osiris, whom she
helped restore to life. Companion and assistant to Isis.

NUN (NU)   The god of primeval water, source of the creation of the
earth.

NUT   The wife of Geb and the mother of Osiris, Isis, Seth,
and Nephthys. She is considered the personification of the sky. She and
Geb were considered to be the givers of food. She is sometimes represented as a female figure, arched over the earth. The sun passes through her
body and is swallowed up in the west, to reappear in the east the next day.

OSIRIS   God of earth and vegetation. He was murdered by his
brother Seth, his body chopped up and scattered over Egypt. He symbolized the yearly drought in his death and the periodic flooding of the Nile
and the growth of grain in his rebirth. Represented holding the crossed
staff and whip, symbols of power and authority.

PTAH   A local god of Memphis, he was the considered the god
of craftsmen. Supposedly he said the names of all the things in the world,
causing them to come into existence. Also known as "the opener of the
day."

RA (RE)   The sun god of Heliopolis. Probably the oldest god
worshiped in Egypt. In early thought, he was the supreme deity, creator
of mankind. Originally known as Temu, god of the setting sun, or "closer of the day."

SETH (SET)   Brother to Osiris, god of the night, also regarded as the Lord of Upper Egypt. Killed his brother Osiris. He was associated with the desert and storms. Portrayed as a animal with big ears resembling a donkey. Later vanquished by Horus, an event symbolizing the triumph of day over night and life over death.

SHU   Son of Atum (later Ra) and Tefnut, goddess of moisture. He made his way between the gods Geb and Nut and raised up Nut to form the sky.

SOBEK   Worshiped in cities closely linked to the Nile. Sobek was depicted as a man with the head of a crocodile.

TEFNUT   Goddess of moisture and twin sister to Shu.

THOTH   The god of wisdom, Thoth advised Osiris. Known for his sagacity and cunning. Also associated with the moon. Usually depicted as a dog-headed ape, but also with the head of an ibis or baboon. Seen seated on top of the balance in the "weighing of the heart."

Sources: Casson (1965), 184–86; Budge (1987), 108–35; Edwards (1993), 7.

UNITS OF MEASUREMENT AND OTHER TECHNICAL DATA

Units of Measure and Conversion Factors

MULTIPLY	BY	TO GET
Centimeters	0.394	Inches
Cubic meters	1.308	Cubic yards
Cubits	0.01	*Khets*
Cubits	20.62	Inches
Cubits	52.4	Centimeters
Digits	0.7364	Inches
Digits	1.871	Centimeters
Digits	0.03571	*Cubits*
Digits	0.025	*Palms*
Feet	0.305	Meters
Grams/cubic centimeter	1	Metric tons/cubic meter
Hekats	0.1358	Bushels
Hekats	4.785	Liters
Kilograms	2.205	Pounds
Kilometers	19.08	*Khets*
Kilometers	0.621	Miles
Meters	1.908	*Cubits*
Meters	3.281	Feet
Meters	39.37	Inches
Metric tons	1.102	Short tons
Newtons	0.2248	Pounds (force)
Palms	0.1429	*Cubits*
Palms	7.486	Centimeters
Stats	10,000	Square cubits
Stats	2,745	Square meters

Italics indicate ancient Egyptian measures.

Angle Measurements

WHEN x IS:	ANGLE OF SLOPE IS: (DEGREES)	EXAMPLES
2 cubits (56 digits)	26.56	Khufu corridors
1.7 cubits (47.5 digits)	30.50	Queen's Chamber ceiling
30 digits	43.03	
29 $^2/_3$ digits	43.34	Red Pyramid (Dahshur)
28 digits (7 palms)	45.00	
26 digits	47.12	
24 digits	49.40	Senwosret I (Lisht)
22 digits	51.84	Khufu (Giza)
21 digits	53.13	Khafre (Giza), Userkaf (Saqqara), Teti (Saqqara), Neferirkare (Abusir), Pepi I (Saqqara), Merene (Saqqara)
20 digits (5 palms)	54.46	Bent Pyramid (Dahshur), Amenemhet I (Lisht)
18 $^2/_3$ digits	56.31	Unas (Saqqara), Senwosret III (Dahshur)
18 digits	57.26	Amenemhet III (Dahshur)
16 digits (4 palms)	60.26	
14 digits	63.43	

x = horizontal setback of sloping face (seqed) when y = vertical rise of 1 cubit (7 palms, or 28 digits).

Properties of Materials

PROPERTY	MATERIAL				
	LIMESTONE	GRANITE	SOIL	COPPER	CEDAR
Density (grams/cm3)	2.21	2.72	1.8	8.8	0.54
ULTIMATE STRESSES (PSI)					
Compression	9,000	20,000			2,000
Shear	1,000	2,300			620
Modulus of rupture	1,200	1,600			4,200
Modulus of elasticity	8,400,000	7,500,000			640,000
WORKING STRESSES (PSI)					
Compression	800	1,200			600
Tension	125	150			
Shear	150	200			

Properties vary widely. These are the average values as used in this book.

CALCULATING THE NUMBER
OF BLOCKS IN THE PYRAMID

COURSE No.	ACTUAL LENGTH PER SIDE (M)	AREA (SQ M)	BLOCK HEIGHT (M)	CUMULATIVE HEIGHT (M)	BLOCK LENGTH (M)
1	230.430	53,098	1.48	1.48	2.30
2	228.402	52,168	1.29	2.8	1.94
3	226.536	51,318	1.19	4.0	2.37
4	224.769	50,521	1.12	5.1	2.25
5	223.191	49,814	1.00	6.1	2.01
6	221.642	49,125	0.99	7.1	1.97
7	219.979	48,391	1.06	8.1	2.12
8	218.528	47,754	0.92	9.1	1.85
9	217.119	47,141	0.90	9.9	1.79
10	215.648	46,504	0.94	10.9	1.87
11	214.324	45,935	0.84	11.7	1.68
12	213.053	45,392	0.81	12.5	1.62
13	211.945	44,921	0.70	13.2	1.41
14	210.775	44,426	0.74	14.0	1.49
15	209.628	43,944	0.73	14.7	1.46
16	208.49	43,468	0.72	15.4	1.45
17	207.366	43,001	0.72	16.2	1.43
18	206.125	42,487	0.79	16.9	1.58
19	204.584	41,855	0.98	17.9	1.96
20	203.634	41,467	0.60	18.5	1.21

Source: Petrie (1883).

Block Width (m)	Net Blocks Needed	Cumulative Blocks	Individual Block Weight (kg)	No. of Casing Stones	Casing Stone Weight (kg)
1.96	6,313	6,313	14,656	470	3,231
1.72	8,404	14,717	9,429	532	2,273
1.58	7,116	21,833	9,798	574	1,822
1.49	7,744	29,576	8,309	601	1,459
1.34	9,513	39,089	5,933	668	1,143
1.31	9,635	48,725	5,601	676	1,181
1.41	8,078	56,802	6,929	625	1,188
1.23	10,431	67,234	4,606	712	879
1.19	10,800	78,034	4,218	728	865
1.24	19,269	97,303	4,799	693	848
1.12	23,598	120,901	3,493	766	659
1.08	25,288	146,189	3,098	792	530
0.94	33,093	179,282	2,049	904	425
0.99	29,309	208,591	2,412	852	465
0.97	30,122	238,713	2,279	863	444
0.96	30,324	269,037	2,220	866	431
0.95	30,754	299,791	2,139	872	464
1.05	24,818	324,609	2,885	785	703
1.30	15,747	340,357	5,515	628	668
0.80	41,651	382,007	1,293	1,013	247

A PRIMER ON
PROGRAM MANAGEMENT

DEFINITION

Program management is the science and practice of managing large private and public works projects. Typically these involve complex engineering design and construction utilizing multiple contractors. The logistical issues—making certain everything comes together at the right time, in the right quantity, with the right quality—is one of the great challenges of these projects and becomes the major preoccupation of the program manager. Because the expenditure of public or private funds is involved, it is often necessary to engage an independent third party to represent and protect the interests of the owner.

CONCEPTS

Any complex project involves the interrelationships of numerous "players": designers, contractors, suppliers, regulators, unions, customers, the public, government officials, and the project owner, to name a few. To achieve the best result for the least cost there invariably will be one optimal way of planning and implementing the work. Normally the optimal approach is not obvious on a complex, multiyear project. In addition, changes can disrupt the otherwise orderly flow of the work—a flood or other natural disaster, a shortage of a critical component or material, an accident, delays in funding, and so on. There is also a logical sequence to performing the work, which ensures that it will be completed in a timely, cost-effective manner. For example, while certain activities must be performed in sequence (forms must be built before foundations can be poured), other activities can go on in parallel (structural steel can be fabricated off-site while foundations are being poured).

Several decades ago it was both time consuming and labor intensive to track the various activities of a major project. This effort was performed by "clerks of the work," who kept records. Anticipation of problems and corrective action depended heavily on the judgment and experience of a field construction superintendent. Often he or she made decisions based on partial or incorrect information, because that was all that was available.

This situation began to change with the advent of digital computers, which permitted the tracking of complex operations through large databases. Then someone realized that the database could incorporate logic and that the computer could be used to simulate the various activities of a project to see how long one approach might take compared to another. With the recent advances in inexpensive and powerful microcomputers it is now possible to locate this tool at the job site, where the program manager uses it and other techniques to keep complex undertakings on track.

The main benefit to the owner is that he or she has an independent professional support staff to assist and oversee complex projects. Since the program manager is not normally involved in design or construction, he or she is able to represent the owner's interests without bias. This system usually improves coordination, reduces the cost of the construction, saves time, and reduces claims that might otherwise arise.

What the Program Manager Does

Normally a project is divided into several phases: preconstruction, design, procurement, construction, and postconstruction. It is preferable to have the program manager begin work during the preconstruction phase, but this is not always done.

If the program manager is brought into the project at the preconstruction phase, the owner benefits. The program manager can assist in preparing a project master schedule and master budget so the owner takes control of the project from the outset, rather than have the schedule determined by the convenience of the designer or the contractor. In another important activity that begins during the design phase of the project, the program manager conducts a series of "constructability reviews," which ensure that the project can be constructed as designed and that the proposed approach is the best one, given local conditions, available materials, budgets, schedule, etc. Prior to construction—during the procurement phase—the program manager will assist the owner with the preparation of bid packages, prequalification of subcontractors, bid evaluation, and the award of contracts. It may be necessary to prepare project manuals and procedures for consistency in the work. Once construction contracts are awarded, the program manager will revise the project master schedule and master budget if necessary and update them on a monthly basis. Once construction starts, the program manager will be responsible for inspection and testing, and sometimes surveying, on behalf of the owner. He or she will establish a document control system to record and archive the thousands of documents that are necessary for payment and other purposes. The program manager also ensures that the contractors implement their quality assurance and safety programs. When construction is complete, he or she will witness start-up and operational tests, review operations and maintenance manuals and as-built drawings, assess liquidated damages if applicable, and approve final payment requests and prepare project closeout documents. These tasks can be summarized as follows.

Main Tasks for the Program Manager

Preconstruction Phase
- Project management plan and procedures
- Management information system
- Work breakdown structure
- Project milesone schedule
- Project master schedule
- Project master budget
- Document control system
- Special technical studies if needed (master planning, site evaluation, infrastructure, etc.)

Design Phase

- Contract administration
- Design management procedures
- Design reviews
- Constructability reviews
- Value engineering reviews

Procurement Phase

- Contracting plan
- Procurement support
- Bidder's conferences, prebid meetings, etc.
- Bid evaluation and award procedures

Construction Phase

- Weekly progress meetings
- Construction management plan
- Resident engineer's manual
- Construction safety manual
- Procedures for requests for information (RFIs), shop drawings, and submittals
- Construction observation, testing, and inspection
- Change order and claims management
- Test start-up plan

Postconstruction Phase

- Prepare project punchlist
- Coordinate and witness systems start-up, testing, and training (commissioning)
- Verify operations and maintenance manuals and warranty information
- Approve final invoices for payment
- Disputes/claims resolution
- Contract closeout

In all of these activities, the program manager functions as an extension of the owner's staff, or as the owner's representative. The program manager's goals are to ensure that contractors meet the schedule, cost, quality, and safety requirements of the owner. The program manager also advises the owner on the payment of contractor's bills and evaluates claims.

The Program Manager's Tools

Essential to the program manager's work is a series of tools that make possible the planning and tracking of complex activities that are being performed simultaneously. Proprietary systems and many commercially available computer programs are used for this purpose today. Most of them employ a series of "modules" that are used for each of the program manager's tasks. Some flexibility with these modules is required, since some owners want the program manager's system to interface and communicate with their own project information system.

Just as important as the tools are the data entered into the program manager's project control system. Clearly, the system is only as good as the accuracy and completeness of the data it contains.

The third key element is personnel. Ultimately, the quality, expertise, and experience of the program management team will determine the success of the effort.

Work Breakdown Structure

The work breakdown structure is the subdivision of the work into all of the individual elements required to complete the project. It is normally prepared in a series of levels, each more complex than the preceding and each involving a greater amount of detail. Level 0 is normally the entire project, while Level 1 is the major subdivisions (such as site acquisition, engineering design, construction, procurement, and so on). Level 2 consists of the subelements of work for each of the Level 1 items. For example, under design there might be civil, structural, architectural, mechanical, and electrical; under construction there might be demolition, site preparation, civil works, structures, and so on. Level 3 goes to the next level of detail—for example, under electrical design there might be power systems, lighting, instrumentation and controls, telecommunication, and so on.

LOGIC DIAGRAM

The logic diagram shows the logical sequence and interrelationships of the different parts of the work. Examples are the logic ties between engineering design, procurement, construction, start-up, acceptance testing, and warranty support, as well as the ties between the different elements of each of these, such as surveying, grading, excavation, and the other components of the civil work.

CRITICAL PATH SCHEDULING

The project schedule incorporates information from the work breakdown structure and the logic diagram to establish a baseline schedule for doing the work. Using such commercially available programs such as Primavera Project Planner, the program manager's project controls staff will develop the schedule and then determine the critical path. The critical path consists of all of the elements of work that influence the project completion date. In other words, tasks not on the critical path can be delayed or extended without delaying the project. The only way to accelerate the work, however, is to shorten the critical path. This can be done by shortening the completion time of items on the critical path or by changing the logic of the work to move a lengthy task off the critical path.

The current practice is to utilize resource-loaded critical path schedules for major projects. These schedules include the resources required to accomplish each task in the project data base. Thus we not only know how long a particular task will take, but also how many labor-hours are planned to be expended, and the quantity (and value) of materials to be put in place during that task. This is a powerful tool for securing accurate appraisals of work completed and work in progress, as well as for evaluating contractor claims and payment requests, particularly if they are based on the percentage of work complete.

COST CONTROL AND ESTIMATING

Cost control refers to tracking project costs, comparing them to project budgets, and evaluating and explaining variances. Cost control can also be used to forecast cash flow requirements for payments. Various systems are used for cost estimating. In some cases it is merely a conventional "spread sheet" with experience data for a given region and type of construction. The database is continuously updated with current data to provide accurate and timely information. Commercial programs as Timberline and other software are also available for cost estimating.

There are several different types of cost estimates. In one case, an estimator "takes off" quantities from drawings and specifications, determines the current prices, and compiles an estimate. This method does not work for certain types of projects, or for situations in which complete designs are not available—design/build jobs and renovation of existing buildings, for example. In these cases a much greater skill level is required because the cost estimator must have experience visualizing and recalling all elements of the work. We call this "conceptual cost estimating."

DOCUMENT CONTROL

A major project involves thousands of documents. Most of these are critical, and if improperly handled can lead to additional expense to the owner. For example, the contractor may submit requests for information (RFIs) to the designer to clarify the designer's intent. If these requests are not handled promptly, the contractor may have to stop work on a particular task while waiting for information. It is usual to establish a system by which every document is logged and tracked and its receipt and eventual disposition is recorded. Documents requiring follow-up action are flagged in periodic reports. The document control system also interfaces with the cost and schedule modules. For example, if a change to the work is submitted, the resulting impact on the program budget and schedule are included in the cost forecast and in the next update of the project schedule and critical path.

Management Information System

Some means of collecting, recording, and providing access to all of the essential information is needed in large public works projects, so the program manager assumes responsibility for establishing and maintaining a management information system.

This system incorporates budget and cost information from the master budget; schedule information; correspondence logs; status reports on safety, quality, and requests for information; project review schedules; and meeting minutes. The document control process may be incorporated into this system. The system can be equipped with a "firewall" that limits access to certain levels of information, depending on the requestor's need for the information. The trend today is to install the management information system on an Internet site so that all concerned parties can access the information. On major public works projects, portions of the information are even accessible to the general public. Examples include a description of the project, percentage complete, and public information bulletins (for example, notice of road closures). Passwords and various security measures control access to sensitive or private information such as workers' salaries.

Change Management

Contractor claims for additional compensation may result in contract changes that require additional compensation or additional time to perform the work. Change orders can arise for several different reasons: a request by the owner for additional work; a delay caused by some event outside the control of the owner or the contractor, such as a storm; unforeseen circumstances, such as poor soil conditions; forced acceleration; an error on the part of the designer or the contractor, which causes some of the work to be redone or replaced; or regulatory or code changes.

In each of these situations, the owner and the contractor will have certain claims to make upon each other. If we assume that both parties act in good faith, then if a change is requested by the owner, he or she will be prepared to pay for it and extend the schedule if necessary to complete the additional work required. In many situations, the full extent of responsibility for each party is not totally clear. For example, the contractor may have performed part of the work incorrectly and be faced with a need to make changes to correct the problem. At the same time, the owner may recognize this correction as an opportunity to improve or modify the work and therefore request that other improvements or modifications be made. In this situation each party now bears some responsibility for the added cost and the delay, and the only way to resolve each party's responsibility is through negotiation. This is the actual situation more often than not, so the program manager will establish a formal process for handling claims.

Typically, after the contractor submits a request for a change, the program manager evaluates it and determines which category it falls into and whether or not it is justified. This may involve modification of the contractor's request in terms of the work that he will do, the payment that he/she will receive, or the schedule required to implement the change. After evaluating the request for change, the owner authorizes the contractor to proceed. This authorization will ultimately take the form of a change order to the contract, which provides the contractor additional funds and direction for the performance of the work.

HEMIUNU AS PROGRAM MANAGER

From the foregoing discussion it should be clear that a major, multiyear project, involving a significant portion of the population of an entire country and extensive use of natural resources and materials, must of necessity have had some way to manage the program. In ancient Egypt, these tasks, or some variation on them, had to have been completed to ensure a successful outcome.

On Hemiunu's team, scribes were the clerks of the work, keeping the essential payroll records on clay tablets or papyri that have long since been lost. Worker attendance, requisitions of stone blocks from the quarries, and other essential information had to be recorded and communicated in some manner. There is considerable evidence suggesting that the ancient Egyptians used program management techniques. For example, large stone blocks have identification numbers written on them, indicating they were cut to size and installed in accordance with a prearranged plan. The progress of the work, including future activities, had to be known to make sure the requisite number of laborers was on site at the right time, or conversely, when laborers were ready to work, that the right amount of materials were available for them. In addition, job administration records from later dynasties have been found. They include daily attendance, payroll information, and other data.

Maintaining the schedule was important, because the project faced an irrevocable but uncertain end date: the pyramid had to be completed before the death of the pharaoh. Certainly no modern program manager has faced an assignment as challenging as this!

NOTES

INTRODUCTION
1. Edwards (1993).
2. Arts and Entertainment Channel (1998).
3. Smith (1999), 34–41.
4. Educational Broadcasting Corporation (2001).

CHAPTER 1
My intent in this chapter is to provide a background for the reader to understand the social and cultural context in which Khufu's pyramid was constructed. For the reader who wants to explore the Old Kingdom in greater detail, I recommend Shaw (2000), Durant (1954), Casson (1965), White (1963), Wilson (1956), and Aldred (1965). I have drawn heavily on these works.
1. Lehner (1997), 7, says the population of Egypt at the time the Giza pyramids were built was 1.6 million.
2. Shaw (2000), 90.
Attributed to Jaromir Malek, the author of Chapter 5, "The Old Kingdom."
3. Durant (1954), 145.
4. Starr (1983), 7.
5. Shaw (2000), 17.
6. Ibid., 21–25.
7. Starr (1983), 11–12.
8. Burns (1955), 8.
9. Shaw (2000), 25–30.
10. Burns (1955), 12.
11. Shaw (2000), 32–35.

12. White (1963), 26; Durant (1954), 145–46.
13. Starr (1983), 55; Durant (1954), 146.
14. Shaw (2000), 39–41.
15. Ibid., 64.
16. Ibid., 47–59.
17. Lehner (1997), 13.
18. Ibid., 75.
19. Andreu (1997), 13.
20. Personal communication, Antonio Loprieno, professor of Egyptology and chair, Department of Near Eastern Languages and Culture, UCLA, August 1999.
21. Casson (1965), 147.
22. Burns (1955), 34.
23. Casson (1965), 31.
24. Shaw (2000), 69.
25. Burns (1955), 35–36.
26. The vizier served as the pharaoh's prime minister, with broad authority and power. Casson (1965), 93–94, provides a brief overview. Andreu (1997), 17, outlines the significance of this position. Van den Boorn (1988), 309–31, has a detailed listing of the vizier's duties.
27. White (1963), 151.
28. Shaw (2000), 77–78.
29. Egyptian religion is a complex subject about which volumes have been written. For greater detail on this subject, Edwards (1993), 6–18, is a brief synopsis; Casson (1965), Chapter 4, is a concise overview; Budge (1987) is a reprint of the classic 1900 treatment

by the late Sir Wallis Budge, of the British Museum; Frankfort (2000) and Cerny (1979), more comprehensive treatments, are cited by many authors. For the Egyptian version of the creation I have followed Edwards (1993), 7; see also Lehner (1997), 34.

30. Budge (1987), 19–22.

31. Budge (1987), 120; Casson (1965), 90–91; Edwards (1993), 6–11.

32. Lehner (1997), 31.

33. Edwards (1993), 179.

34. Andreu (1997), 144.

35. Budge (1987), 161–65; White (1963), 75–76. See Casson (1965), 81–89, for an interesting photo essay.

36. Budge (1987), 186; Casson (1965), 76.

37. Budge (1987), 189–91; and Romano (1990), 3–8, discuss the significance of the ba and the ka.

38. Siliotti (1997), 12.

39. Lehner (1997), 28–30.

40. Ibid., 18–19.

41. White (1963), 162–64.

42. Casson (1965), 30–32.

43. Ibid., 100–101.

44. Ibid., 12–13.

45. Ibid., 32.

46. White (1963), 60.

47. Ibid., 134.

48. Ibid., 144.

49. Hayes (1964), 158–59; Casson (1965), 125–29.

50. Continental Granite Corporation (1990). This company manufactures replicas of granite royal cubit rods.

51. Hayes (1964), 162–63; White (1963), 65–66.

52. Hayes (1964), 108–9; White (1963), 70.

53. White (1963), 108–11; Casson (1965), 33; Lehner (1997), 119.

54. Lehner (1997), 117.

55. Two limestone models of pyramids, one corresponding to the Step

Pyramid, the other representing King Amenemhet III's pyramid at Hawara, are described in Edwards (1993), 260.

56. Shaw (2000), 95.

57. I drew from various sources in constructing the family tree, including Shaw (2000), 94–98; Lehner (1997), 116, 120, 122; and the Web site of the Royal Ontario Museum (www.rom.on.ca/egypt/case/society/who). This Web site had brief biographical sketches of approximately 100 persons: pharaohs, their wives, their children, their viziers, and priests and other notables from the Third through Sixth Dynasties.

CHAPTER 2

The ancient Egyptians built more than 100 known pyramids along the western banks of the Nile River. Lehner (1997) has an excellent compilation of the most significant ones. My purpose here is to examine four of the early pyramids that I believe influenced the design of Khufu's pyramid. Just as the Step Pyramid was a natural extension of experience gained from building ever larger mastabas, Khufu's pyramid was made possible by the trial-and-error learning experiences gained by building these earlier structures.

1. Gillispie and Dewachter (1987), vol. 5, *Memphis et environs*, plate 2, no page number.

2. Lehner (1997), 87.

3. Siliotti (1997), 122–23.

4. Lehner (1997), 103.

5. Personal communication, Mark Lehner, Dahshur, February 11, 1997.

6. Carl Sandburg, "Cool Tombs," in West (1926), 138.

7. Siliotti (1997), 41; Clarke and Engelbach (1990), 217.

8. Clarke and Engelbach (1990), 217.

9.   Clarke and Engelbach (1990), Chapter 20, 216–23, has an excellent overview of Egyptian mathematics.

10.   Arnold (1991), 8.

11.   Clarke and Engelbach (1990), 46–48.

12.   Edwards (1993), 246.

13.   Lehner (1997), 212.

14.   Edwards (1993), 99.

15.   Clarke and Engelbach (1990), 67.

16.   Edwards (1993), 250–51.

17.   Brinker and Taylor (1958), 267.

18.   Novokshchenov (1996), 51.

19.   Clarke and Engelbach (1990), 207.

20.   Arnold (1991), 268–69, provides a synopsis of data on ancient ropes. He states that ropes were made from palm fibers, flax, reeds, papyrus, or halfa grass. They were made by twisting fibers into yarns and making ropes of three to five strands. Examples are known with diameters of 6.3 to 6.8 centimeters, while calculations indicate there may have been ropes with diameters as large as 18.4 centimeters and working loads of 6.5 metric tons.

21.   Shaw (2000), 41.

22.   Ibid., 55, 58–59.

23.   Ibid., 105, 114. An elaborate drainage system at Sahure's pyramid (Fifth Dynasty) employed more than 300 meters of copper pipe, including copper-lined basins (Edwards 1993, 166–67).

24.   Lehner (1997), 210.

25.   Clarke and Engelbach (1990), 17–18, 202–4.

26.   Ibid., Chapter 4, 32–45.

CHAPTER 3

Khufu's pyramid is also referred to as the Great Pyramid or the Pyramid of Cheops (the name the Greeks gave to Khufu). Lehner (1997), Edwards (1993), and Siliotti (1997) have excellent general descriptions of the main features. Siliotti's book has excellent photographs and a section written by Zahi Hawass. For detailed measurements, the classic works are Vyse (1840) and Petrie (1883), which contain compilations of measurements of every imaginable feature of the monuments at Giza.

1.   Thomas (1995), 125; Lehner (1997), 8.

2.   Lehner (1997), 126.

3.   Edwards (1993), 245.

4.   Herodotus (1972), 179.

5.   Today the Internet provides a forum for many interesting (and in some cases bizarre) theories concerning the pyramids. Go to www.google.com, search for "Great Pyramid Construction," and you'll get 136,000 entries: www.catchpenny.org will take you to "Catchpenny Mysteries of Ancient Egypt" and a section entitled "Alternate Theories of Pyramid Construction," along with some discussion of why the alternate theories don't make sense. You can read a theory that the pyramid is made of cast-in-place concrete; various lever theories; levitation theories in which the pyramids were massive antennas that tapped into a mysterious energy field; or theories that the pyramids were built by a lost civilization in Atlantis or interplanetary travelers. Then, there is the Williams Hydraulic Theory, which postulates that the stones were lifted by a giant hydraulic system that ran vertically up the center of the pyramid. Or see, for example, Cray (1999), 8, for experiments on using kites to lift stone obelisks.

6.   Edwards (1993), 102.

7.   Ibid., 104.

8.   Ibid., 104–5.

9. Ibid., 106. See also Vyse (1840) and Petrie (1883) for additional dimensions and measurements.

10. Although the height and width could be measured, it was not always possible to determine how long each stone was.

11. Petrie (1883). See Chapter 6, "Outside of Great Pyramid," and plate 8, "Courses of the Great Pyramid."

12. According to Petrie's data, the courses where the variation in elevation fell within this range were 4, 9, 19, 22, 24–25, 30, 32–34, 37–39, 45, 49–51, 59, 67 69, 78–80, 83–85, 93–94, 107–10, 126–28, 131–32, 176, 184, 193, and 198 (Petrie 1883, plate 8).

13. No one can state with certainty how many blocks of stone are in the pyramid, since complete details of the interior construction are not known. I believe that the ancient Egyptians used any available odd shapes to fill in the inner areas to minimize the labor needed to cut stones. However, my model assumes uniform sizes. I do deduct the volume of void spaces occupied by all of the corridors and chambers and make an allowance for the rock outcropping incorporated into the base of the pyramid.

14. Lehner (1997), 116.

15. Williams (1953), 306.

16. Lehner (1985), 45–50.

17. Edwards (1993), 112–13.

CHAPTER 4

After gaining an appreciation for the organizational and administrative skills of the ancient Egyptians, it seemed logical to me to use the concept of program management as a framework for analyzing the design and construction of Khufu's pyramid. In this construct, every project has its program manager.

For example, in the renovation and rebuilding of the damaged Pentagon, there was the genius of Walker Lee Evey on behalf of the government, and from my team, equally dedicated people—Les Hunkele and Stacie Condrell (see Winston [2002] or Hunkele, Sabbatini, and Murph [2001] for the dramatic details). I settled on Hemiunu as the de facto program manager for building the pyramid, although in the interest of telling the story, I may have credited him with more day-to-day involvement than was actually the case.

1. Lehner (1985b), 113, has a vision of the Giza Plateau as it might have been before construction started (see fig. 3a).

2. Ibid., 43.

3. Ibid., 126–27.

4. Ibid., figure 3c, 127; Lehner (1997), 204–5; McClintock (2001), 44.

5. Hadingham (1992), 40; Barnes (1996), 70.

6. Petrie (1883), 43.

7. Lehner (1997), 218; Arnold (1991), 11–13.

8. Herz-Fischer (2000), 168.

9. Recently, archaeologists have identified various dates and quarry marks on stones in the Red Pyramid. From these, one can infer that about one-fifth of the pyramid was completed in two years (Verner [1998], 185).

10. Andreu (1997), 40–41; and Edwards (1993), 248–51, describe the ritual of stretching the cord and marking the corners. Fifth Dynasty tomb paintings show the pharaoh, accompanied by priests, holding a stake with an attached cord in one hand and a mallet in the other. Edwards describes the *merkhet*, used to tell time by star elevations at night or sun shadows during

the day; and the *bay*, a piece of wood with an aperture used to sight distant objects.

CHAPTER 5

Recent work by Zahi Hawass and Mark Lehner has shed light on the lives of the people who lived at Giza and constructed the pyramids. Their findings confirm the existence of a construction camp, workshops, and a commissary at the site. Just readying the site for construction was a mammoth task, and one that has not been addressed at all by previous writers. For this reason, I have gone into some detail in describing the various steps that were necessary before construction of the pyramid could begin.

1. David (1996). See Chapter 3, "The Towns of the Royal Workmen," 56–98, for an excellent description of conditions in three ancient workers' villages. I have used this work to imagine what the workers' village at Giza was like.
2. Ibid., 61–67.
3. Lehner (1997), 236–37.
4. Lehner, personal communication, Giza, Egypt, February 9, 1997.
5. Hawass and Lehner (1997), 39–43.
6. Hawass (1992), 67. Hawass breaks down the settlement area into three components: the "pyramid city," which supported the administration of the pyramid complex and housed priests and nobility (3,000 persons); the artisans' village, housing the skilled workforce of 5,000 persons; and a workers' village, where 10,000 laborers lived in huts. These ruins appear to encompass 2 square kilometers. Today they reflect an evolution of at least seventy years, through the end of Menkaure's reign and possibly longer, so they must have been somewhat less in size and scope during Khufu's reign.
7. Lehner (1997), 225. Lehner believes that two crews of 2,000 skilled workers could have accomplished the main pyramid building tasks (excluding laborers for constructing the ramps, carpenters for various tasks, toolmakers, cooks, bakers, potters, etc.).
8. Bierbrier (1982), 50–54; also see Hagen and Hagen (1999), 74–93, for a chapter on "the first workers' strike."
9. Arnold (1991), 268.
10. David (1996), 242.
11. Casson (1965), 32.
12. David (1996), 217.
13. Personal communication, Dan Eberle, June 30, 2003. Eberle is a program manager for AECOM Government Services, Inc., stationed in the Sinai Desert in support of the MultiNational Force and Observers.
14. Sisler, Vanderwerf, and Davidson (1949), 782.
15. David (1996), 216.
16. Ibid., 221–25.
17. Levy et al. (2001), 23–28.
18. Aldred (1965), 59.
19. Casson (1965), 128.
20. Lehner (1997), 13.
21. Edwards (1993), 245.
22. Partridge (1996), 8.
23. Hawass, personal communication, February 12, 1997. In my field notes I noted that Mark Lehner thought that the massif rose to a height of 7 meters in the pyramid.
24. Lehner (1997), 212–13.
25. Mark Lehner, personal communication, Giza, Egypt, February 11, 1997. He discussed the method of cutting out the stones on the north side of Khafre's pyramid while taking me on a tour of the Giza site.
26. Edwards (1993), 246, describes

the concept of leveling with water. Regarding the leveling of the site, he states that the southeast corner is 1.25 cm higher than the northwest corner.

27. Arnold (1991), 27–52.

28. Edwards (1993), 102.

29. Lehner (1997), 111.

30. Ibid., 212.

31. Houdin and Houdin (2002), 5–6.

CHAPTER 6

The foremost question in most people's minds when they look at the Great Pyramid is, "How did the Egyptians lift all of those heavy stones?" The evidence for the use of ramps is compelling, but the *exact* form of the Giza ramps is uncertain. The answer to this question no doubt can be found at the site, but it awaits further exploration. The ancient Egyptians used various forms of ramps, and no single method seemed to be preferred. Instead, the approach was adapted to the needs of the site or the experience of the builder. For this reason, I have devoted an entire chapter to the subject of ramps, analyzed the options, and selected an approach—as ancient Egyptian engineers might have done—subject to the same constraints regarding slope, building methods, and materials evident in the ruins of ancient ramps.

1. Regarding the use of bearing stones as substitutes for pulleys, see Arnold (1991), 283; and Lehner (1997), 211.

2. For written records of ramp calculations, see Clarke and Engelbach (1990), 92; and Edwards (1993), 262.

3. Arnold (1991), 97; Clarke and Engelbach (1990), 92.

4. Arnold (1991), 96; Clarke and Engelbach (1990), 93.

5. Arnold (1991), 85, 86, 90.

6. Ibid., 82–83.

7. Lehner (1997), 215–17, describes remains of ramps at Meidum, the Sinki pyramid at south Abydos, and Sekhemkhet's pyramid at Saqqara. Remains of ramps and transport roads are found at Lisht; one ramp sloped at 8° (Lehner 1997, 226–27). Edwards (1993), 262, cites examples of ramps and ramp retaining walls made of brick at the temple of Karnak and the mortuary temple of Mycerinus. Arnold (1991), 78–101, provides a detailed discussion of ramps. At Giza, I observed two examples. One is the retaining wall of a large ramp southeast of Khufu's pyramid, leading toward the third of the Queens' Pyramids (called G1c by archaeologists). This ramp may have been used in the construction of the Queens' Pyramids. The other leads toward a mastaba in the western cemetery.

8. Hawass (1990), 42.

9. Hawass (2003), 4.

10. Hodgman (1959), 2121, gives the density of limestone as 2.68–2.76 grams per cubic centimeter. Note that this is numerically equivalent to 2.68–2.76 metric tons per cubic meter, which is the unit I use throughout the book. Arnold (1991), 28, gives the density of "porous" limestone as 1.7–2.6 metric tons per cubic meter, and that of "dense" limestone as 2.65–2.85 metric tons per cubic meter. He further states that the density of granite is 2.6–3.2 metric tons per cubic meter, and that of Aswan granite, 2.68 metric tons per cubic meter.

11. Petrie (1883), Chapter 7; Vyse (1840), 109–14. I used Petrie's and Vyse's measurements of the dimensions of the internal corridors and chambers, converting them to SI units

and rounding. I also compared the main chamber measurements to dimensions reported in Edwards (1993), Lehner (1997), and Hawass (1990), and spot-checked against some limited measurements I made myself.

12. For typical slopes, see Arnold (1991), 79–97; Edwards (1993), 262, 271; and Lehner (1997), 216.

13. Partridge (1996), 132–33.

CHAPTER 7

This chapter is devoted to the actual construction of the pyramid and includes a description of how I believe the principal internal features were erected. Here I have to beg the reader's indulgence in a disconcerting switch in measurement units—from meters to cubits. If you will bend your arm at the elbow and extend your middle finger, the distance from the elbow to the tip of your finger is approximately one cubit (perhaps exactly, if you carry the blood of some ancient pharaoh in your veins!). When we understand that the King's Chamber is precisely 10 cubits wide and 20 cubits long (the Queen's Chamber is 10 cubits wide and 11 cubits long), and they are placed respectively 82 and 41 cubits above the base, suddenly the logic and planning of the builders becomes more apparent than when we try to make sense out of a room that is 5.3 meters wide and 10.4 meters long. I hope that the insertion of cubit measurements in this chapter will give the reader an appreciation for the Egyptian designers' sense of proportion and beauty.

1. Petrie (1883). The dimensions of corridors and chambers in this chapter have been extracted from Petrie's classic work.

2. Lehner (1997), 206.

3. Ibid., 224. See also Edwards (1993), 255–56; Clarke and Engelbach, 21; and Arnold (1991), 61.

4. Smith (2001), 46–47.

5. Arnold (1991), 282–83.

6. Houdin and Houdin (2002); Hovland (2003).

7. The basic dimensions are all from Petrie (1883). I converted selected measurements to cubits because it gives insight into the designers' ideas (1 cubit = 52.4 centimeters).

8. Lehner (1997), 112.

9. Petrie (1883), 69.

10. Arnold (1991), 179–83.

11. Petrie (1883), 72.

12. Petrie (1883), 76–77.

13. Ibid., 80.

14. Mencken (1963), 29.

15. Hovland (2003), 15–17.

16. Hovland (2003), 13, figure 7.

17. Petrie (1883), 92–93.

18. Lehner (1997), 16–17.

CHAPTER 8

Another central question in the minds of many who visit Giza is, "How many people did it take to build the pyramid?" To address this question, I broke down the project into the major elements of work and then determined how much labor it took to perform each element. In any construction project the job is staffed to meet the schedule, and the level of staffing varies as major tasks are completed. Workers finish one part of the job and move on to the next part, and this was certainly the case in the construction of the pyramid. I cite this to illustrate that some judgment is involved in preparing these estimates, and another builder might approach it with different assumptions. Also, one has to set a limit on the level of detail to be included: I have consciously left out some tasks that I know were done (painting, for example), simply because there is not enough information available to even guess at the scope of the work. But there is enough contingency in the overall numbers to cover the items I ignored.

1. Arnold (1991), 63.
2. Edwards (1993), 256–57.
3. Arnold (1991), 72.
4. Ibid., 215. A photograph shows Eleventh Dynasty workers' baskets, still filled with soil and rubble, abandoned at a temple construction site.
5. Merriman (1930), 769–70. One man can loosen 0.4–0.8 cubic meters per hour in hard soil with a pick or mattock, or 2.3–3.8 cubic meters per hour in ordinary loam. A laborer can shovel earth into a wagon at the rate of 0.9–1.1 cubic meters per hour. Combining picking and loading, a laborer can do about 0.8 cubic meters of loam per hour, 0.6 cubic meters per hour of stiff clay, and 0.4 cubic meters per hour of hardpan. Clarke and Engelbach (1990), 91, estimate that a large hall could be filled with earth at the rate of about 1.3 cubic meters per labor-hour.
6. Herodotus (1972), 178–79.
7. Arnold (1991), 66.
8. Lehner (1997), 119.
9. Hovland (2003), 5–7. Hovland has kindly shared his unpublished manuscript with me. In it, he carries out a thorough and interesting analysis of static and kinetic coefficients of friction for various materials (wood on wood, wood on sand or clay, etc.) in both dry and lubricated (wet) conditions. His analyses yield friction coefficients ranging from 0.15 to 0.70. He also back-calculated that the friction coefficient for the famous Djehutihotep statue (see page 186; and Lehner 1997, 203) is 0.11. My choice of 0.5 is probably conservative, assuming the hauling surfaces were moistened, but I elected to use a higher value to compensate for other factors not considered in my analysis. For example, once the sledge was in position, the block of stone was levered off the sledge onto the next course, where the friction coefficient of stone on stone was higher.
10. Stewart (1998), 98–100. I am indebted to Ian Stewart for this idea. He reports on work performed by S. T. Weir, who performed a similar analysis. I used the model I had developed with Petrie's course-measurement data and slightly different assumptions. Weir assumed that construction went on for 365 days per year for twenty-three years and concluded that a smaller average workforce could do the job. Merriman (1930), 143–44, provides a table indicating the amount of work that can be accomplished by a laborer in an eight-hour day. Examples: carrying a load upstairs, returning unloaded, 542 kilojoules per day, average power output, 18.8 watts; wheeling earth in a barrow up a 1:12 slope, returning unloaded, 483 kilojoules per day, 16.8 watts; hammering, 650 kilojoules per day, 22.6 watts. I reduced the average of these values by about 50 percent and rounded it to 10 watts to arrive at 288 kilojoules per day.
11. Lehner (1997), 211.
12. Hawass and Lehner (1997), 31.
13. Lehner (1997), 231.
14. Hawass (1997b), 39–43.
15. Hawass and Lehner (1997), 31–38.
16. Hawass (1997b), 43.
17. Morell (2001), 81, cites an inscription in the tomb of Kai about the workers who built it: "I paid them

in beer and bread, and made them make an oath they were satisfied."

18. Siliotti (1997), 96–97.

19. Lehner (1997), 230.

20. Bierbrier (1982), 54.

21. Morley (1991). Temple records from Kahun, Twelfth Dynasty, specify eight jugs of beer and sixteen loaves of bread as the daily ration.

22. Jones and Jones (1982), 15–23. Nile Valley Egyptians invented the oven and found that a mixture of wheat and water left in a warm place would ferment and rise, or leaven, producing a lighter bread. They made more than thirty kinds of bread. Early forms were made from barley and emmer wheat. Breads contain 60 to 70 percent flour by weight, and 30 to 40 percent liquid and other ingredients. Initially the liquid fraction is higher, but 10 to 13 percent of the raw weight is lost during baking. From this I estimated that a 2-kilogram loaf is 65 percent flour. Beer is commonly made from malted barley, but wheat or other grains can be used. Malting is the process whereby the grain is mixed with water and allowed to sprout, creating sugar and enzymes. When the sprouts are still very small, the grain is dried and roasted. Brown ale is 12.5 percent malt by weight, and wheat beer is 18.5 percent (Lutzen and Stevens [1994], 69, 178). A 2-liter jug would be 15 percent grain.

CHAPTER 9

Another major question concerning the construction of the pyramid is, How long did it take? Fortunately, one of the major contributions of program management has been the introduction of new computer tools for analyzing project schedules. If a project can be broken down into a series of logical activities, and if the sequence of these activities is known or can be determined, and if the labor to perform each activity is known or can be estimated, accurate determinations of the time to complete the work can be made. Another benefit of these tools is that they enable the program manager to evaluate the consequences of project changes. For example, what is the impact on the schedule of adding labor or changing the sequence of the work? Now it should be apparent that the sequencing of Chapters 5 through 8 was to develop the information needed for Chapter 9.

1. Lehner (1997), 109.

2. Edwards (1993), 124.

3. El-Baz (1988), 51.

CHAPTER 10

Were the pyramids successful? Why did the Egyptians reach the height of the Khufu's Great Pyramid, and then never achieve it again? Did building mammoth masonry structures become a lost art? There is an ancient Arab saying: "All the world fears time, but time fears the pyramids." I suppose we can say that the pyramids have been victorious in their battle with time so far, but time has had its effects on the pyramids, and time is limitless and will continue to degrade these great structures. For this reason, preservation needs to receive more attention.

1. I borrowed this thought from Kadare (1996), a novel about the palace intrigues that might have accompanied the construction of the pyramid.

2. Shaw (2000), 109. The tomb of Neferirkara (Fifth Dynasty) contained a set of papyri that described the day-

to-day operations of a pyramid complex: administrative records of produce delivered to the city, inventories of goods, lists of priests who were on duty, letters, and other details. We may surmise that a similar administrative system once existed at Giza.

3.　An illustration in McClintock (2001), 44–45, shows what this walled city might have looked like, based on Lehner's excavations.

4.　Maugh (1993), 1. According to Hawass, the pharaoh gave everyone (referring to the pyramid workers) permission to build their tombs any way they liked.

5.　Shaw (2000), 105, 116–17, discusses the economic impacts of these policies and the economic decline that marked the end of the Old Kingdom.

6.　Hemiunu's statue (minus the eyes) now resides in the Roemer-and-Pelizaeus Museum, Hildesheim, Germany.

7.　Peck (1997), 31. Hemiunu's tomb, now known as mastaba G 4000, was excavated in 1912 by Hermann Junker.

8.　Cartonnage, a material made from layers of cloth, plaster, glue, and paint, was often used to cover a mummy.

9.　Khufu's mummy and coffin have never been found, so this description is purely speculative. It is consistent, however, with descriptions of royal mummies from later periods that were found intact.

10.　Although Khufu's mummy never has been found, the sarcophagus is damaged, indicating it was opened at some time in the distant past; the plugs in the Ascending Corridor were released, blocking the corridor; and the external opening was indeed sealed.

11.　Lehner (1997), 120–21.

12.　Ibid., 121.

13.　Siliotti (1997) includes a description of the boat pit opened in 1954. Hawass (1990), 25, states that the boat pit contained eighteen cartouches with Djedefre's name. El-Baz (1988), 513–20, describes the discovery of the second boat.

14.　The Heisenberg Uncertainty Principle, named for physicist Werner Heisenberg, states that the very act of making measurements on subatomic particles changes them, because interaction is required to detect them and interaction changes their physical state.

15.　Smith and Peoples (2000), 5–10, describes methods and elements of a site management plan.

16.　A large-scale geographical information system has been developed to record and retrieve photographs of artifacts and other data from an extensive archaeological dig, along with the spatial coordinates where objects are found (Levy et al. [2001], 23–29).

17.　Remote sensing with a cesium magnetometer is used to identify mud-brick walls and sand foundations of buildings buried beneath Nile clay (Pusch 1999). The National Aeronautics and Space Administration has developed remote-sensing techniques using radar, infrared, or ultraviolet sensing to evaluate global climate change, agriculture, and environmental pollution (Smith and Peoples [2000], 9). The same techniques have revealed long-lost canal beds, ancient roadways, and trails.

GLOSSARY

alcove. A large recess in a room, generally separated from the main area by an arch.

architrave. A main beam, supported by columns; in classical architecture, part of an entablature supporting a decorative frieze.

backfill. Soil or other material used to replace material removed during construction; also the process of replacing material excavated from trenches and foundations.

back shoot. In surveying, to make a sighting back to a previous measurement point.

backing stones. Stones placed accurately on the perimeter of the pyramid to control its dimensions and shape.

balk (baulk). A large, roughly squared timber, placed on the ground as a support. Also called a sleeper.

beam. The width of a vessel at its widest point.

bedrock. Solid rock that underlies surface soil and is used to support foundations.

benchmark. A point where the elevation of the ground or the work is known and used as a reference in determining the elevation or height of other points.

cartonnage. A material made from layers of cloth, plaster, glue, and paint, and used to cover mummies or make face masks for them.

cartouche. An oblong or oval shape enclosing the name of the pharaoh (carved or painted).

casing stones. The outer layer of stones installed on a pyramid to give it its smooth, sloping finish.

causeway.   A paved roadway, usually elevated, over uneven terrain or water.

clear and grub.   To remove tree stumps, shrubs, roots, or debris before excavation and leveling of a site.

construction gap.   An area of construction where the work is deliberately left unfinished to install interior features, after which the gap is closed.

corbelling.   Courses of masonry where each successive course overlaps the preceding one to form a self-supporting projection.

core stones.   The innermost stones used in constructing the pyramid. They did not need to be cut with precision dimensions, as did the backing stones and casing stones.

critical path.   The specific series of activities in a critical path schedule that actually control the completion date of the overall project. If any activity on the critical path is delayed, the project completion date is extended. Delays in completing activities not on the critical path will not change the project completion date unless their float is exceeded.

critical path schedule.   A scheduling, planning, and project control methodology whereby a project is broken down into a series of activities, the duration of each activity is determined, and all activities are linked in a logical manner with their predecessor and successor activities to determine the total time to complete the work.

cubit.   From the Latin cubitum, meaning "elbow." The Egyptian measure of length, from the elbow to the tip of the middle finger; in this book, 52.4 centimeters.

cut and fill.   To remove soil from the high part of a sloping site and place it on the lower portion for leveling, or to cut roads or canals and use the excavated material for embankments or leveling.

declination.   The angular distance above or below the celestial equator; used to measure the position of the sun, other planets, or stars.

digit.   A unit of measurement, a subdivision of the cubit: 28 digits equal a cubit, and 4 digits equal a palm.

dip.   The slope of layers of soil.

displacement.   The carrying capacity of a vessel; literally, the amount of water displaced by the vessel.

dolerite.   A granular, crystalline igneous rock, very hard, used by the ancient Egyptians to cut and shape granite.

escarpment.   A steep slope, often separating two different geological formations, resulting from erosion or faulting.

falsework.   A temporary structure erected to support work being constructed, typically consisting of supporting columns, beams, lateral bracing, etc.

float.   In a critical path schedule, extra time available to complete a specified activity without affecting the completion date of the overall project.

flux.   In metal production, an additive that aids in the removal of impurities.

grade.   The slope of a road or natural ground; the finished surface of a building site, roadbed, or canal bed.

hekat.   A unit of volume used by the ancient Egyptians, equal to 4.785 liters or 0.1358 bushels.

Hemiunu.   Cousin to Khufu and his vizier; considered the person responsible for seeing the Great Pyramid built for Khufu.

Khafre.   Son of Khufu, the fourth pharaoh of the Fourth Dynasty, builder of the middle pyramid at Giza.

khet.   A unit of length, equivalent to 100 cubits.

Khufu.   The second pharaoh of the Fourth Dynasty in ancient Egypt, 2551–2528 BC; responsible for the Great Pyramid at Giza.

lintel.   A beam of wood, stone, or steel placed horizontally across the top of a door or window opening to support the wall above the opening.

logic diagram.   A schematic representation of the relationship between the various activities that make up a large complex project, i.e., which activity or activities must be completed before another can be started.

massif.   In geology, a rock mass, usually a block or mountain produced by displacement or faulting.

Memphis.   Capital of ancient Egypt.

Menkaure.   Son of Khafre, the fifth pharaoh of the Fourth Dynasty, and builder of the third large pyramid at Giza.

Neolithic.   Related to the latest part of the Stone Age; characterized by polished stone implements.

neutron activation analysis.   The technique of irradiating a material in a nuclear reactor so its component elements become radioactive and its composition can be readily determined by gamma ray spectroscopy. This highly sensitive method can detect parts per billion of trace elements.

nome.   From the Greek, *nomos*, meaning "district"; a province of ancient Egypt.

offset.   In surveying, a horizontal distance measured perpendicular to a known horizontal (or vertical) line.

overburden.   The layer of soil that covers bedrock or a mineral deposit.

Paleolithic.   Related to the second part of the Stone Age; characterized by chipped-stone implements.

palm.   A unit of measurement, a subdivision of the cubit; 7 palms equal a cubit.

phyle.   An ancient Egyptian work gang; derived from the Greek word for tribe; also the largest political subdivision of the ancient Athenians.

plumb bob.   A pointed weight hung from a string and used for vertical alignment of walls and columns.

portcullis.   A lifting gate at the entrance of a stronghold or protected room; in medieval times, with sharp spikes at the bottom. In ancient Egypt, the gates were slabs of granite.

Predynastic period.   In ancient Egypt, roughly 5000 to 3000 BC, before the time of the pharaohs.

preliminary design.   The early development of a set of architectural plans and specifications, prepared for the owner's review prior to detailed or final design.

program management.   The science and practice of managing large private and public works projects in an economical and timely manner.

pyramidion.   The capstone, or topmost stone of a pyramid.

quay.   A paved area next to a harbor, used for loading and unloading vessels.

rod and chain.   Used in surveying and leveling, a marked post and a surveyor's steel tape.

saddle roof.   A gabled pitched roof. A pointed saddle roof is one in which the roof beams meet in the center and form an upside-down V.

seqed.   The ancient Egyptian term for measuring slope or angle of inclination. (Refer to Appendix 2.)

stat.   A unit of area used by the ancient Egyptians, equal to 10,000 square cubits.

talfa.   A calcareous clay material derived from Nile River mud, used by the ancient Egyptians for various building purposes, including mud bricks and (when moistened) lubrication for stone hauling.

trepanning.   To remove a circular section of the skull for treating brain injuries.

turning basin.   In a harbor, an area set aside for maneuvering vessels.

White Wall.   Name given to the royal residence in ancient Memphis; synonymous with the capital of ancient Egypt.

work breakdown structure.   The result obtained by dividing a complex project into a series of individual steps, called "activities," each of which must be completed to complete the entire project.

Definitions checked in *Webster's New Collegiate Dictionary* (1981) and *Construction Dictionary* (National Association of Women in Construction, 1989).

ANNOTATED BIBLIOGRAPHY

Aldred, Cyril. 1965. *Egypt to the End of the Old Kingdom*. New York: McGraw Hill. Culture, arts, and religion; also concepts for construction, including four ramps, one at each corner.

Andreu, Guillemete. 1997. *Egypt in the Age of the Pyramids*. Translated by David Lorton. London: Cornell University Press. Religion, quarries, construction sequence and ramps, science and mathematics, lives of workers.

Arnold, Dieter. 1991. *Building in Egypt: Pharaonic Stone Masonry*. Oxford: Oxford University Press. Excellent details on building methods, materials, and tools. One of the best and most thorough descriptions of Egyptian masonry construction methods.

Arts and Entertainment Channel. 1998. *The Great Builders of Egypt*. Produced by Greystone Productions for the A&E Network.

Bierbrier, Morris. 1982. *The Tomb Builders of the Pharaohs*. London: British Museum, A Colonnade Book. Village life at Deir el-Medina, where workmen constructed the royal tombs. Work gangs, methods, records, etc.

Brinker, Russell C., and Warren C. Taylor. 1958. *Elementary Surveying*. 3rd ed. Scranton, PA.: International Textbook Co.

Budge, E. A. Wallis. 1967. *The Egyptian Book of the Dead: The Papyrus of Ani*. Mineola, NY: Dover Publications. First published in 1895, this is the classic translation of *The Book of the Dead*, with extensive notes and explanations.

———. 1987. *Egyptian Religion*. Secaucus, NJ: Carol Publishing Group, Citadel Press. A somewhat dated but definitive description.

Burns, Edward M. 1955. *Western Civilizations: Their History and Their Culture*. New York: W. W. Norton.

Casson, Lionel. 1965. *Ancient Egypt*. New York: Time-Life Books.

Cerny, Jaroslav. 1979. *Ancient Egyptian Religion*. 1952. Reprint, Westport, CT: Greenwood Press. An excellent, concise, and very readable survey.

Clarke, Somers, and R. Engelbach. 1990. *Ancient Egyptian Construction and Architecture*. New York: Dover Publications. A classic work on Egyptian masonry, rock cutting, mortar, construction methods, tools, design, decoration, construction, and mathematics, with extensive illustrations.

Continental Granite Corporation. 1990. *The Royal Cubits*. This manual accompanies a replica of a cubit rod sold by the company. It describes the history of the cubit and the Egyptian measuring system. Good bibliography.

Cray, Dan. 1999. "How Do You Build a Pyramid? Go Fly a Kite." *Time*, December 6.

David, Ann Rosalie. 1996. *The Pyramid Builders of Ancient Egypt: A Modern Investigation of the Pharaoh's Workforce*. New York: Routledge. A detailed study of the workmen who built tombs and pyramids during the Twelfth Dynasty (Kahun) and Eighteenth to Twentieth Dynasties (Tell el-Amarna and Deir el-Medina).

Durant, Will. 1954. *The Story of Civilization*. Part 1, *Our Oriental Heritage*. New York: Simon and Schuster.

Educational Broadcasting Corporation. 2001. *The Secrets of the Pharaohs*. Part 2, *Lost City of the Pyramids*. New York: Public Broadcasting Service.

Edwards, I. E. S. 1993. *The Pyramids of Egypt*. Baltimore: Penguin Books. A classic description of Egyptian pyramids. The revised edition incorporates more recent findings.

El-Baz, Farouk. 1988. "Finding a Pharaoh's Funeral Bark." *National Geographic*, April, 513–50. The discovery of the funeral boats at Giza.

Frankfort, Henri. 2000. *Ancient Egyptian Religion: An Interpretation*. 1948. Reprint, New York: Dover Publications. A classic originally published by Columbia University Press.

Gillispie, Charles C., and Michel Dewachter, eds. 1987. *The Monuments of Egypt*. 1809. Reprint, Old Saybrook, CT: Konecky & Konecky. Comprising the complete archaeological plates from Napoleon's expedition to Egypt, this work was originally printed in ten volumes as *La description de l'Egypte* (Paris: Imprimerie Imperiale, 1809).

Hadingham, Evan. 1992. "Pyramid Schemes." *Atlantic Monthly*, November, 38–52. This summary of Mark Lehner's and Zahi Hawass's research in the Giza Plateau also has a summary of Lehner's PBS *Nova* film.

———. 1997. "Now, On the Count of Three, Everybody Pull." *Smithsonian*, January, 22–32. Experiments to determine the labor required to raise a 40-ton obelisk.

Hagen, Rose-Marie, and Rainer Hagen. 1999. *Egypt. People. Gods. Pharaohs.* Cologne, Germany: Benedikt Taschen. A "coffee table" book with excellent photos and descriptions of history, politics, religion, and culture.

Hawass, Zahi A. 1990. *The Pyramids of Ancient Egypt.* Pittsburgh: Carnegie Museum of Natural History. An excellent summary of the Giza pyramids, including construction methods, materials, and the function of the pyramid complex.

———. 1992. "The Workmen's Community at Giza." In *Haus und Palast im Alten Agypten*, Internationales Symposium, April, Cairo.

———. 1997a. "The Red, White, and Black." *Horus Magazine* (Egypt Air), January–March, 9–14, 24–28. Pyramids at Abusir and Dahshur, and a description of Deir el-Medina.

———. 1997b. "Tombs of the Pyramid Builders." *Archaeology*, January–February, 39–43. Worker's tombs, characteristics of the work-force, health of workers, etc.

———. 2000. "Development of the Royal Mortuary Complex." Web site, guardians.net/hawass/mortuary. An extensive description of the Giza site, with notes on building pyramids, workmen, etc.

———. 2003. "Pyramid Construction: New Evidence Discovered at Giza." Web site, guardians.net/hawass/pbuildrs. The location of the ramp and quarry.

Hawass, Zahi A., and Mark Lehner. 1994. "The Sphinx: Who Built It and Why." *Archaeology*, September–October, 30–47.

———. 1997. "Builders of the Pyramids." *Archaeology*, January–February, 31–38. Workers' village, workshops, bakeries, harbor, and other facilities that may have existed during the pyramid construction.

Hawass, Zahi A., Mark Lehner, Shawki Nakhla, Georges Bonani, Willy Wölfli, Herbert Haas, Robert Wenke, John Nolan, and Wilma Wetterstrom. 1999. "Dating the Pyramids." *Archaeology,* September–October, 27–33. New radiocarbon studies indicate that the average date of the age of Khufu was 2694 BC, about a century earlier than previously thought.

Hayes, William C. "Egypt." 1964. In *Everyday Life in Ancient Times: Highlights of the Beginnings of Civilization in Mesopotamia, Egypt, Greece, and Rome,* edited by Gilbert Grosvenor. Washington, DC: National Geographic Society. This volume contains over 100 paintings of events in the daily life of ancient civilizations.

Herodotus. 1972. *The Histories.* Translated by Aubrey de Selincourt. London: Penguin Books. Herodotus speculated on the number of workers and methods used to construct the pyramids.

Herz-Fischer, Roger. 2000. *The Shape of the Great Pyramid.* Waterloo, Ontario: Wilfrid Laurier University Press. Debunks many of the quack mathematical theories about the angles, areas, and dimensions of the Great Pyramid.

Hodgman, Charles D., ed. 1959. *Handbook of Chemistry and Physics.* 41st ed. Cleveland: Chemical Rubber Publishing.

Houdin, Jean-Pierre, and Henri Houdin. 2002. "La construction de la Grande Pyramide. La seule methode plausible." *Traveaux* 792 (December). A proposed system of counterweights and internal ramps.

Hovland, H. John. 2003. "Construction of the Pyramids at Giza: Use of Sleds, Ropes, and Ramps." Manuscript.

Hunkele, Lester M., III, Julian Sabbatini, and John Murph. 2001. "The Pentagon Project." *Civil Engineering,* June, 38–45, 87, 106–7.

Jones, Judith, and Evans Jones. 1982. *The Book of Bread.* New York: Harper & Row.

Kadare, Ismail. 1996. *The Pyramid.* New York: Arcade Publishing. Originally published in Albanian and translated from the French, this is a fictional account of Khufu and Hemiunu, and treason in high places in ancient Egypt.

Lehner, Mark. 1985a. "The Giza Plateau Mapping Project: Season 1984–85." *American Research Center in Egypt Newsletter* 131 (fall): 23–56. Detailed mapping of the site and features around the base of Khufu's pyramid to determine if holes were for surveying and to establish baselines, elevations, etc.

———. 1985b. "The Development of the Giza Necropolis: The Khufu Project." In *Mitteilungen des Deutschen Archäologischen Instituts Abteilung Kairo*. Mainz am Rhein, Germany: Philipp Von Zabern, 109–46. Interesting presentation of Lehner's concept of the Giza site development, quarries, ramps, etc.

———. 1985c. "The Pyramid Tomb of Hetep-heres and the Satellite Pyramid of Khufu." Mainz am Rhein, Germany: Phillip von Zabern. Photographs of marker holes at Khufu and Khafre's pyramids, and further explanation of the survey and leveling process.

———. 1986. "The Giza Plateau Mapping Project." *American Research Center in Egypt Newsletter* 135 (fall): 29–54. Detailed mapping of the site, and mapping of features around the base of Khafre's pyramid.

———. 1992. *This Old Pyramid*. PBS *Nova* video. WGBH Educational Foundation. Egyptologist Mark Lehner and stonemason Roger Hopkins reconstruct a small version of the Great Pyramid.

———. 1996. "The Pyramid." In *Secrets of Lost Empires: Reconstructing the Glories of Ages Past*, by Michael Barnes et al. London: BBC Books. Good description of finding north; data from Lehner (1992); ramps and quarry operations.

———. 1997. *The Complete Pyramids*. New York: Thames and Hudson. A "whole earth catalog" of pyramids in Egypt.

Lehner, Mark, and David Goodman. 1996. "New Mapping of Ancient Egypt." *P.O.B. Magazine* (surveying trade journal), February, 16–24. Giza survey methodology.

Levy, Thomas E., James D. Anderson, Mark Waggoner, Neil Smith, Adolfo Muniz, and Russell B. Adams. 2001. "Interface: Archaeology and Technology." *SAA Archaeological Record* 1, no. 3 (May): 23–29. A huge complex of copper smelters and copper-tool manufacturing shops in the Jabal Hamrat Fidan region of Jordan. The site dates to 2300–2600 BC and is unique because it was destroyed by an earthquake, effectively preserving intact the tools, casting molds, and artifacts.

Lutzen, Karl, and Mark Stevens. 1994. *Homebrew Favorites*. Pownal, VT: Storey Publishing.

Maugh, Thomas H. 1993. "Tombs Give Glimpse of Life at Pyramids." *Los Angeles Times*, August 30. This article on research by Zahi Hawass, director of Giza Plateau for Egyptian Antiquities Organization, summarizes discoveries in the workers' tombs at Giza.

McClintock, Jack. 2001. "Lost City." *Discover Magazine*, October, 40–47, 88. An article about the workers' village and Mark Lehner. See the reconstruction of the village on page 44 and references on page 88.

Mencken, August. 1963. *Designing and Building the Great Pyramid*. Baltimore: privately printed by Schneidereith & Sons. This small book has some engineering insights into the construction of the pyramid.

Merriman, Thaddeus, ed. 1930. *American Civil Engineers' Handbook*. 5th ed. New York: John Wiley & Sons.

Morell, Virginia. 2001. "The Pyramid Builders." *National Geographic*, November, 79–99. An overview of Mark Lehner's work and Zahi Hawass's exploration of the workers' tombs. Data on health of workers, food, etc.

Morley, J. 1991. *An Egyptian Pyramid*. New York: Peter Bedrick Books.

Novokshchenov, Vladimir. 1996. "Pyramid Power." *Civil Engineering*, November, 50–53. Supports the idea that pyramids were built with local limestone. Describes mortar and stone characteristics. Gives limestone density as 2.1 grams/cubic centimeter.

Partridge, Robert. 1996. *Transport in Ancient Egypt*. New York: Rubicon Press. Boats, river navigation, and wooden sledges.

Peck, William H. 1997. *Splendors of Ancient Egypt*. New York: Detroit Institute of Arts, Abbeville Press. Published in conjunction with an exhibition of art and artifacts from the Roemer and Pelizaeus Museum, Hildesheim, Germany, arranged by the Florida International Museum, Saint Petersburg.

Petrie, W. M. Flinders. 1883. *The Pyramids and Temples of Gizeh*. New York: Scribner and Wellford.

———. 1990. *The Pyramids and Temples of Gizeh*. Edited by Zahi Hawass. Rev. ed. London: Histories and Mysteries of Man. An abridged version of Petrie's seminal work, with an update on recent discoveries by Zahi Hawass.

Pusch, Edgar. 1999. "Toward a Map of Piramesse." *Egyptian Archaeology* 14:20–23. The use of a cesium magnetometer to reveal mud-brick walls beneath agricultural fields.

Romano, James F. 1990. *Death, Burial, and the Afterlife in Ancient Egypt*. Pittsburgh: Carnegie Museum of Natural History.

Shaw, Ian, ed. 2000. *The Oxford History of Ancient Egypt*. Oxford: Oxford University Press.

Siliotti, Alberto. 1997. *Guide to the Pyramids of Egypt*. Vercelli, Italy: Barnes and Noble. Photographic tour of the major pyramids with an introduction and a section on Giza pyramids by Zahi Hawass.

Sisler, Harry H., Calvin A. Vanderwerf, and Arthur W. Davidson. 1949. *General Chemistry: A Systematic Approach*. New York: MacMillan. Refer to Chapter 41 for the basic chemistry of copper production.

Smith, Craig B. 1999. "Program Management B.C." *Civil Engineering*, June, 34–41. An analysis of the labor, time required, and methods for constructing the Great Pyramid.

———. 2001. "How Pyramids Were Built." *Lafayette Magazine*, spring, 46–47. A synopsis of the Judith A. Resnik Memorial Lecture, presented by the author at Lafayette College, Easton, PA, in October 2000.

Smith, Craig B., and Ann Peoples. 2000. "Site Management at Archeological Monuments." *Proceedings, Eighth International Congress of Egyptologists*, Cairo, April.

Starr, Chester G. 1983. *A History of the Ancient World*. 3rd ed. Oxford: Oxford University Press.

Stewart, Ian. 1998. "Mathematical Recreations: Counting the Pyramid Builders." *Scientific American*, September, 98–99. An analysis of the labor required to build the pyramid. Uses measures of the useful work of a man and the potential energy of stone placement to estimate workforce.

Stierlin, Henri. 1992. *The Pharaohs: Master Builders*. Paris: Editions Pierre Terrail. Excellent photographs, examples of the use of grid lines, moving a heavy statue. Beautifully illustrated in color, this book covers the pyramids, tombs, temples, and Egyptian art.

Thomas, Nancy. 1995. *The American Discovery of Ancient Egypt*. Los Angeles: Los Angeles County Museum of Art. The introduction has a concise history of the Old Kingdom. The book has details on Egyptian art but not much information on the pyramids.

van den Boorn, G. P. F. 1988. *The Duties of the Vizier: Civil Administration in the Early New Kingdom*. London: Kegan Paul International. Chapter 3 is a translation of texts on the duties of the vizier.

Verner, Miroslav. 1998. *The Pyramids*. Translated from German by Steven Rendall. New York: Grove Press.

Vyse, Howard. 1840. *Operations Carried on at the Pyramids of Gizeh in 1837*. Vol. 2. London: James Fraser. Has sketches and measurements of the pyramids at that time, with drawings of internal details.

West, Rebecca, ed. 1926. *Selected Poems of Carl Sandburg*. New York: Harcourt, Brace.

White, Jon Manchip. 1963. *Everyday Life in Ancient Egypt*. New York: G. P. Putnam's Sons.

Williams, Oscar, ed. 1953. *The Golden Treasury of the Best Songs and Lyrical Poems*. Rev. ed. New York: New American Library.

Wilson, John A. 1956. *The Culture of Ancient Egypt*. Chicago: University of Chicago Press.

Winston, Sherie. 2002. "Pentagon's Construction Team Beats the Odds on One-Year Rebuild." *Engineering News-Record*, September 2, 6–10.

ILLUSTRATION CREDITS

INDEX

Boldface type indicates illustrations.

Abu Roash, 238, 239
access roads, 88, 109–11, 118, 230, 235
AECOM Technology Corporation, 13. *See also*
    DMJM and note 13, Chapter 5
Akhet Khufu, 124
angles of pyramids. *See* seqed
annual flooding, Nile River. *See* inundation
archaeological site management, 242
architectural drawings, 75, 77
area of a circle, ancient method for computing, 73
Arnold, D., 75, 143, 153–55, 190, 204, 211
Aswan, 39, 108, 111, 116, 118, 120, 122, 137, 144,
    194, 203, 210, 213, 220, 226–27, 232

ba, 43, 45, 67, 115
backing stones, 66, 84, 139, 174, 184, 190–91, 193;
    definition and purpose, 102, 104, 147–48
Badarian, 32
bakeries, 110, 112, 129, 132, 217, 220, 240
barges, 50, 84, 106, 111, 137–38
Bechtel Corporation, 126
bedrock protrusion (massif), 69, 112, 160, 162,
    172, 227
benchmarks, 78, 142, 147, 183, 199
Bent Pyramid, 25, 44, 54–55, 59, 64, 65, 66–68,
    72, 113, 139, 157–58, 202, plate 6
boats, 39, 49, 50, 62, 70, 82, 84, 106, 136, 138, 217,
    232, 238; boat pits, 50, 106, 133, 201, 230,
    232–33, 238
Book of the Dead, The 42
bosses, stone handling, 184, 185
breweries, 110, 112, 129, 132
Byblos, 38, 39, 50, 136

Cairo Marble and Granite Company, 13
calendar, 36–37, 48, 72, 227, 230
capstone. *See* pyramidion
cartouche; Djedefre, 239; Khufu, 195
casing stones, 18, 23, 61–64, 66, 68–69, 97, 100,
    104–5, 119–20, 122–23, 143, 147, 160, 162,
    176–77, 185, 198, 203, 207, 210–11, 213,
    226–77, 234; Menkaure's pyramid, 71, 201;
    Red Pyramid, 66

causeway, 46, 60, 64, 106–7, 110, 112, 119, 201,
    208, 218, 223, 230–33, 239
Cheops. *See* Khufu
Clarke, S., 76, 78, 84, 154
construction gap, 147, 190
construction lines, 76–77, 184
Construction Logic Diagram, 116, 118–19
construction schedule, 120, 121, 122, 153, 164, 172,
    202, 204, 207. *See also* Chapter 9
copper, 19, 31–32, 38–39; ore, 33, 48, 82, 133;
    smelting, 82, 91, 129, 134–36; tools, 19, 31, 33,
    48–49, 82–84, 90–91, 93, 129, 133–36, 144,
    183, 212, 217
corbelled ceiling, 25, 62–69, 95, 97, 188, 192
core blocks, 61, 64, 147, 162, 191, 213
Critical Path Schedule, 221, 223–27, 228–29, 231
Cro-Magnon man, 30–31
cubit rod, 48, 50, 73, 74, 237

Dahshur, 25, 36, 44, 49, 55, 59–60, 64, 86, 108,
    116, 152, 157, 158, 173, 232
Deir el-Medina, 127–28, 130, 218
design, 75–7; floor plan, 75; grid lines, scaling, 76
Djedefre, 51, 70, 238–39
Djehutihotep, 185
Djoser, 36, 44, 51, 54–55, 58, 60, 68, 82, 133
DMJM, 21–22, 24, 126
DMJMH+N. *See* DMJM
dolerite, 82, 84, 144, 210

Edwards, I. E. S., 22, 54, 80, 94, 142, 162
Engelbach, R., 76
evolution; mastaba to pyramid, 44; of pyramid
    design, 58, 59

family tree (Fourth Dynasty), 52–53
foundations, 25, 64–65, 67–69, 78, 84, 89, 92, 94,
    124, 146–48, 175, 224–25, 235
Fourth Dynasty, 26, 35–36, 39, 44, 51, 54–55, 58,
    69, 71–73, 75, 81–82, 86, 116, 133, 144, 177, 186,
    241

Giza site, 18, 28, 56–57, 69, 86, 87, 108–9, 151, 236, plate 3, 18
grading, 110, 111, 139, 142, 211
graffiti, 100, 195
Grand Gallery. See Khufu's pyramid
granite, 17, 20, 33, 39, 60, 62, 70–71, 81–83, 85, 89, 97–98, 100, 111, 118, 120, 122, 137, 144, 153, 166, 172, 180, 184, 190, 192–95, 197, 203–4, 210, 212–13, 221, 231–32, 238

harbor, 22, 88, 108–11, 127, 137, 138, 203, 210, 226–27, 230–31, 235, 241
hauling team, 172, 173, 186, 189, 211
Hawass, Z., 24, 25, 27, 91, 108, 110, 129–30, 139, 175, 205, 214, 216, 242–43
Hemiunu, 23–24, 54, 56–57, 59, 68, 71, 85, 87–89, 92, 107, 110–12, 114, 116, 117, 120, 122–24, 137, 144, 150–53, 157–58, 175–76, 178, 197, 209, 222–23, 232, 234–35, 237–39, plate 12
Herodotus, 90, 206
Hetepheres, 51, 105
Hoover Dam, 17, 88, 89, 126
Horus, 36, 41, 42. See also Appendix 1, Egyptian Gods
Huni, 51, 54, 68

Imhotep, 54, 60, 68
inundation, Nile River, 37, 137

jewelry, 31–32, 39, 46, 82

ka, 42–43, 45, 115
Kahun, 127, 134, 135
Khafre, 18, 46, 55, 70, 71, 86–87, 89, 106, 131–32, 142, 158, 198, 232, 235, 239–40
Khafre's pyramid, 70, 71, 141
Khufu, 16, 18, 20–21, 23, 25, 27–29, 39, 43, 44, 46, 50–51, 54, 55–60, 67–71, 77, 82–86, 88–90, 92, 95, 98, 100, 104–6, 108–10, 112–13, 116, 119, 123–24, 132–34, 136–37, 139, 142, 148, 150–54, 156–58, 160, 177, 179, 188, 195, 198, 201–2, 204, 207–8, 212, 216, 221–24, 226–27, 230, 232–40
Khufu's pyramid, 18, 21, 25, 28, 29, 58, 59, 67–68, 69, 70, 79, 86, 92, 93, 95, 105, 142, 152, 156, 159, 212, 221, 232, 235; plates 1, 11, 15; air shafts, 99; Ascending Corridor, 91, 94–98, 119, 149, 158, 179–80, 187, 190, 192, 194, 227, 238; blocking plugs, 95, 180, 192, 194; Descending Corridor, 91, 92, 93, 94, 97, 106, 118, 147, 158, 178–80, 187, 190, 227, 239; dimensions, 78–79, 86; Grand Gallery, 96, 97, 104, 119, 158, 162, 166, 179–80, 188–90, 192–93, 195–96, 227, 238; King's Chamber, 67, 92, 97, 98, 99,

100–101, 104–5, 111, 119–20, 153, 158, 162, 165, 169, 172, 175, 180–81, 190, 193–95, 197, 227, 239, 241; Lower Chamber, 92–94, 118, 147, 158, 178, 227; mortuary temple, 45, 71, 84, 106–7, 110, 119, 201, 208, 216, 222–23, 226, 230–33, 238; portcullis, 19, 97, 193–94, 240; Queen's Chamber, 92, 95–96, 99, 101, 104, 119, 149, 158, 162, 180–81, 187–90, 192, 194–95, 197, 227, 241, plate 9; Relieving Chambers, 70, 99, 100–101, 104, 119, 166–67, 169, 175, 180–81, 195–97, 227, 241; sectional view, 92; Valley Temple, 46, 106, 110, 112, 119, 201, 216, 223, 230–31, 232–33, 235, 238,

labor estimates, 210–13
Lafayette College, 187
law, 36, 38
Lehner, M., 24–25, 65, 78, 84, 91, 106, 108–10, 129–30, 137, 140, 147–48, 184, 203, 212, 217, 241
leveling, 75, 77–79, 91, 112, 139, 141, 142, 146–47, 183, 191, 203, 213, 227
lever system, for lifting stones, 90–91, 156
limestone, 17, 24, 28, 33, 39, 54, 60, 62, 65–66, 68; size of quarry, 110–11; strength and properties, 81, 160
lumber, 36, 38, 49, 50, 61, 82, 91, 128, 132, 136–37, 155–56, 175, 195, 217, 221

Malek, J., 30
mastaba, 44, 54, 60, 61, 64, 237, 239
mathematics, 17, 36, 39, 48, 73, 153
measurements, 48, 67, 73; instruments, 74
Meidum, 25, 54–55, 59, 62, 63, 64, 156, 202, plates 2, 5; grave robber's mallet, 241
Memphis, 35, 39, 108, 150
Mencken, A., 194
Menes, 33, 35
Menkaure, 18, 51, 55, 71, 77, 86, 104, 183, 200, 212, 232, 235, 240
Menkaure's pyramid, 18–19, 71, 71, 86–87, 212
models, 68; computer, 102, 104, 160, 161; program management, 114; pyramid, 54, 77, 88, 119
Mokkatam formation, 109, 142
mortar, 17, 61, 66, 81, 90–91, 148–49, 174, 182, 183, 198, 217
mud bricks, 48–49, 81, 128, 129, 132, 174, 205; dimensions, 49

Napoleon's Expedition, 60, 62
Naqada, 32–33, 82
Neolithic, 31–32
nomes, 33, 35, 47
North pyramid. See Red Pyramid

Nut, 40–41, plate 10. *See also* Appendix 1, Egyptian Gods

Ockman, S., 13
Old Kingdom, 36, 40, 46–51, 153–54, 177
orientation, 79
Osiris, 40–42, 47

Paleolithic, 30–31
Palm, unit of length, 48. *See also* Appendix 2
papyrus, 42, 46–47, 73, 132, 154, 203, 236
Petrie, W. M. F., 101–2, 103, 112, 159, 181, 193–94
phyle, 214
plan, of Giza site, 87
plumb line, 75, 80
population of Egypt, 28, 206
Predynastic, 30–35, 82, 133
program management, 21–23, 56, 114, 116, 122, 125, 216, 233. *See also* Appendix 4
pyramid texts, 40–41
pyramidion, 66, 102, 104, 159, 176, 198, 234

quarries, 28, 66, 75, 88–91, 100, 108–11, 116, 118–19, 122–23, 127, 129, 139, 143, 144, 145, 146, 154, 160, 172, 177, 179, 183–86, 191, 203–4, 206–10, 216, 226–27, 230, 235
Queens' Pyramids, 86, 105, 106, 154, 232

Ra, 40, 41, 67. *See also* Appendix 1, Egyptian Gods
ramps, 73, 84, 89, 91, 119, 122–23, 148, 151, 153–54, 155, 156–57, 159, 162, 163, 176, 177, 185–86, 188, 203, 205–6, 208, 210–13, 223, 226, 230, 232, 234, plate 16; auxiliary, 164, 166, 167, 168, 169–70, 175, 188, 230; evidence for, 91, 108, 110, 156; internal, 190, 191; main, 149, 163, 164, 165, 168, 170–71, 174, 195–96, 200–201, 216, 230, 232; spiral, 162–63, 168, 196, 198; stair case, 169, 170, 171, 177, 198
Ralph M. Parsons Company, The, 126
Red (North) Pyramid, 25, 65, 66, 122, 140, 158, 178, plates 7, 8
Rhind Papyrus, 73, 154
rollers, 48, 82, 91, 132, 137, 146, 153, 174, 183, 191, 212
rope, 46, 75, 82, 84, 132–33, 146, 153, 184, 187, 196, 199, 212, 216, 221, 236

sailing, 50, 84, 111, 133
Saqqara, 24–25, 36, 39, 44, 49–50, 54, 58–59, 61, 91, 108, 116, 152, 198
sarcophagus, Khufu's pyramid, 83, 85, 98, 118–20, 123, 193, 203, 238, 240; Menkaure's pyramid, 20
scaffolding, 65, 91, 97, 104, 132, 177, 212, plate 13

scale drawings, 48, 76
Scorpion King, 33
Scribes, 38, 46–47, 59, 120, 124, 214–15, 219, 235
sensitivity analysis, 11
seqed, 68, 112, 113, 179
Sinai Desert, 33, 82, 126, 133. *See also* note 13, Chapter 5
sledges, 84, 89, 91, 132, 136–37, 144, 154, 173, 175, 184, 186, 191, 198, 212
Sneferu, 36, 44, 50–51, 54–55, 58–59, 62, 64–66, 68–69, 82, 133, 157–58, 223
Sphinx, 71, 90, 109, 142, 154, 241
Step Pyramid, 25, 36, 39, 44, 54–55, 58–59 60, 61, 72, 91, 239, plate 4
stonecutting, 143–44
Stretching the Cord, 11, 123–24, 139
sun-shadow method, 80, 81
surveying, 72, 75, 77, 79, 110, 124, 140, 183, 193
surveyor's marks, 66; in Menkaure's pyramid, 183, 184

talfa, 156, 186
taxation, 37–38, 47, 72
Third Dynasty, 26, 36, 41, 44, 51, 54, 60
Timber. See lumber
tools, 17, 19–20, 22–23, 28, 30–33, 36, 46–49, 58, 72, 75, 81–82, 83, 88–89, 91, 94, 118, 125, 129, 131–36, 142, 144, 179, 184, 204, 217, 223, 240, 241; hauling basket, 141, 205
Transport, 84, 118–19, 213, 226
Turah, 39, 90, 104, 106, 108, 111, 116, 118, 120, 122, 138, 143, 147, 198, 203, 207, 210, 213, 220, 226–27, 234

units of measurement, 48. *See also* Appendix 2
Upper and Lower Egypt, 33–36, 34, 48, 152

vizier, 23, 38, 54–57, 60, 235, 237, 239

Wall of the Crow, 109, 129, 217, plate 17
Washington Monument, 17
weighing the heart ceremony, 42, plate 14
wheels, 19, 48, 84, 89
White Wall, 35, 150
Wood. See lumber
Work Breakdown Structure, 23, 114, 115, 207, 216
workers; houses, 49, 109, 128, 132, 137, 217–18, 234; medical care, 110, 130, 214, 217; payment, 46, 218–20; tombs, 47, 131, 175, 205, 214, 216; village, 109, 111, 118, 122, 127–28, 129, 134, 137, 139, 217–18; work gangs, 47, 100, 152, 175, 189, 212, 214
Workforce, 129, 202–3, 207, 215–16, 219